Measuring [illegible] Development Expenditures

IN THE U.S. ECONOMY

Panel on Research and Development Statistics at the
National Science Foundation

Lawrence D. Brown, Thomas J. Plewes, and Marisa A. Gerstein,
Editors

Committee on National Statistics

Division of Behavioral and Social Sciences and Education

NATIONAL RESEARCH COUNCIL
OF THE NATIONAL ACADEMIES

THE NATIONAL ACADEMIES PRESS
Washington, D.C.
www.nap.edu

THE NATIONAL ACADEMIES PRESS • 500 FIFTH STREET, N.W. • Washington, DC 20001

NOTICE: The project that is the subject of this report was approved by the Governing Board of the National Research Council, whose members are drawn from the councils of the National Academy of Sciences, the National Academy of Engineering, and the Institute of Medicine. The members of the committee responsible for the report were chosen for their special competences and with regard for appropriate balance.

This study was supported by Contract No. SRS-0244598 between the National Academy of Sciences and the National Science Foundation. Support of the work of the Committee on National Statistics is provided by a consortium of federal agencies through a grant from the National Science Foundation (Number SBR-0112521). Any opinions, findings, conclusions, or recommendations expressed in this publication are those of the author(s) and do not necessarily reflect the views of the organizations or agencies that provided support for the project.

Library of Congress Cataloging-in-Publication Data

Measuring research and development expenditures in the U.S. economy / editors, Lawrence D. Brown, Thomas J. Plewes, Marisa A. Gerstein.
p. cm.
Panel to Review Research and Development Statistics at the National Science Foundation. The panel was convened in January 2002 by the Committee on National Statistics of the National Academies at the request of the National Science Foundation to conduct an in-depth and broad-based study of the Research and Development Statistics program of the NSF Science Resources Statistics (SRS) Division. The study was later mandated by Congress in the National Science Foundation Authorization Act of 2002.
Includes bibliographical references.
ISBN 0-309-09320-1 (pbk.) — ISBN 0-309-54595-1 (pdf)
1. Scientific surveys—United States. 2. Science and industry—United States. 3. Research, Industrial—United States. 4. Research—Economic aspects—United States. 5. Research, Industrial—Economic aspects—United States. 6. Science and industry—Government policy—United States. 7. Research, Industrial—Government policy—United States. I. Brown, Lawrence D. II. Plewes, Thomas J. III. Gerstein, Marisa A. IV. National Research Council (U.S.). Panel to Review Research and Development Statistics at the National Science Foundation.
Q149.U5M32 2004
658.5'7'0973—dc22

2004022518

Additional copies of this report are available from the National Academies Press, 500 Fifth Street, N.W., Lockbox 285, Washington, DC 20055; (800) 624-6242 or (202) 334-3313 (in the Washington metropolitan area); Internet, http://www.nap.edu.

Printed in the United States of America.

Suggested citation: National Research Council. (2005). *Measuring Research and Development Expenditures in the U.S. Economy.* Panel on Research and Development Statistics at the National Science Foundation, Lawrence D. Brown, Thomas J. Plewes, and Marisa A. Gerstein, Editors. Committee on National Statistics, Division of Behavioral and Social Sciences and Education. Washington, DC: The National Academies Press.

THE NATIONAL ACADEMIES

Advisers to the Nation on Science, Engineering, and Medicine

The **National Academy of Sciences** is a private, nonprofit, self-perpetuating society of distinguished scholars engaged in scientific and engineering research, dedicated to the furtherance of science and technology and to their use for the general welfare. Upon the authority of the charter granted to it by the Congress in 1863, the Academy has a mandate that requires it to advise the federal government on scientific and technical matters. Dr. Bruce M. Alberts is president of the National Academy of Sciences.

The **National Academy of Engineering** was established in 1964, under the charter of the National Academy of Sciences, as a parallel organization of outstanding engineers. It is autonomous in its administration and in the selection of its members, sharing with the National Academy of Sciences the responsibility for advising the federal government. The National Academy of Engineering also sponsors engineering programs aimed at meeting national needs, encourages education and research, and recognizes the superior achievements of engineers. Dr. Wm. A. Wulf is president of the National Academy of Engineering.

The **Institute of Medicine** was established in 1970 by the National Academy of Sciences to secure the services of eminent members of appropriate professions in the examination of policy matters pertaining to the health of the public. The Institute acts under the responsibility given to the National Academy of Sciences by its congressional charter to be an adviser to the federal government and, upon its own initiative, to identify issues of medical care, research, and education. Dr. Harvey V. Fineberg is president of the Institute of Medicine.

The **National Research Council** was organized by the National Academy of Sciences in 1916 to associate the broad community of science and technology with the Academy's purposes of furthering knowledge and advising the federal government. Functioning in accordance with general policies determined by the Academy, the Council has become the principal operating agency of both the National Academy of Sciences and the National Academy of Engineering in providing services to the government, the public, and the scientific and engineering communities. The Council is administered jointly by both Academies and the Institute of Medicine. Dr. Bruce M. Alberts and Dr. Wm. A. Wulf are chair and vice chair, respectively, of the National Research Council.

www.national-academies.org

PANEL ON RESEARCH AND DEVELOPMENT STATISTICS AT THE NATIONAL SCIENCE FOUNDATION

LAWRENCE D. BROWN (*Chair*), Department of Statistics, The Wharton School, University of Pennsylvania
JOHN L. ADAMS, RAND Corporation, Santa Monica, California
WESLEY M. COHEN, Fuqua School of Business, Duke University
FRED GAULT, Science, Innovation, and Electronic Information Division, Statistics Canada, Ottawa
JAY HAKES, Jimmy Carter Library and Museum, Atlanta
BRONWYN H. HALL, Department of Economics, University of California, Berkeley
CHRISTOPHER T. HILL, Public Policy and Technology, George Mason University, Fairfax, VA
STEVEN KLEPPER, Department of Economics and Social Science, Carnegie Mellon University
JOSHUA LERNER, Department of Finance and Department of Entrepreneurial Management, Harvard Business School
BARUCH LEV, Stern School of Business, New York University
GARY McDONALD, Department of Mathematics and Statistics, Oakland University, Rochester, MI
NORA CATE SCHAEFFER, Department of Sociology, University of Wisconsin, Madison
RICHARD VALLIANT, Joint Program in Statistical Methodology, University of Michigan

THOMAS J. PLEWES, *Study Director*
MARISA A. GERSTEIN, *Research Assistant*
TANYA M. LEE, *Program Assistant*
BARBARA A. BAILAR, *Consultant*
MICHAEL McGEARY, *Consultant*
WILLIAM F. LONG, *Consultant*

Contents

Preface

The nation's commitment to research and development (R&D) has a history that goes back to the earliest days of the United States; some date that commitment to the patent clause in the Constitution, which set the stage for an interplay between the government and the scientific community that continues to this day. Scientific discoveries, technical applications, and, more recently, federal R&D programs have been integral to the nation's development. The role of R&D in the U.S. economy has been characterized as essential to national prosperity. The statement "Research and development widely recognized as being key to economic growth" begins the discussion of trends in R&D expenditures in the National Science Board's report, *Science and Engineering Indicators—2002* (p. 4). The report goes on to suggest that R&D decision making is critical to the future of the U.S. economy and national well-being.

Recognizing the importance of research and development to national prosperity, the United States and other countries have developed systems that conceptualize, define, and measure research and development activities so as to understand and to quantify the influence of R&D on the economy. It is claimed that the level of expenditures and the composition of those expenditures in R&D may be considered a proxy measure of national and sectoral commitment to economic growth and development, and that expenditures are an indication of the "perceived economic importance of R&D relative to all other economic activities" (National Science Board, 2002:4).

Despite this high level of attention to the role of R&D in the economy, the U.S. government devotes little of its treasury to the measurement of

R&D. If the relative spending of the federal government on the measurement of R&D expenditures can be viewed as a measure of its commitment to R&D, that commitment is problematic. In fiscal year 2003, the federal government spent approximately $4.2 million on the National Science Foundation programs to measure R&D expenditures, a minuscule part of the nearly $4.7 billion national investment in statistical programs.

The fast-changing environment for research and development severely impacts the R&D Statistics Program managed by the Science Resources Statistics Division (SRS) at the National Science Foundation (NSF). The NSF mission includes being a source of information for policy formation by other agencies of the federal government and providing data and analysis on a broad policy area for public- and private-sector constituents. SRS products also inform congressional policy makers and NSF itself.

As a major organization in the National Science Foundation, SRS has a requirement for a 5-year review of its programs by an NSF Committee of Visitors.

The review requirement has been partially met in the recent past by such activities as a U.S. Census Bureau advisory committee that recently reviewed the statistical foundation of the R&D industry survey; an academic R&D advisory panel that contributed guidance on collection of information on R&D activities in colleges and universities; and a series of workshops, such as a federal agency R&D administrator workshop in 1998, that have been periodically convened to examine certain aspects of program operations. The R&D surveys have also been reviewed in several recent studies by the National Research Council (2000, 2001c) and also by both the U.S. General Accounting Office (2001) and the Congressional Research Service (1999).

These reviews identified problems with specific R&D surveys, with inconsistencies in data among surveys, and with the inability of the current portfolio to address important policy questions and research issues generated by changes in how R&D is carried out in the United States today. Examples of these problems include high nonresponse rates to items in the industrial R&D survey; discrepancies of several billion dollars between the amount of money that federal agencies reported as R&D support and the amount that the performers of the R&D work reported spending; inconsistent classification of R&D by field across surveys; and the failure of existing surveys to capture interdisciplinary research and collaborations among firms, between research institutions, and across sectors.

In view of the special role of SRS, NSF leadership decided that a review of the R&D statistics programs by a panel of the National Research Council's Committee on National Statistics (CNSTAT) would meet the internal requirement for a periodic program review. A review of the entire portfolio of R&D data collection activities of SRS is timely to identify its strengths and weaknesses and to recommend ways to improve data quality and relevance. In addition to fulfilling the requirement for a periodic outside evaluation, the NSF leadership sponsored this review as a means of developing a comprehensive, integrated review of concepts, definitions, survey methodology, and dissemination.

As result of this interest, the Committee on National Statistics convened the Panel on Research and Development at the National Science Foundation in early 2003 to conduct an in-depth and broad-based study to look at the Research and Development Statistics Program of the Science Resources Statistics Division. The goal of the panel has been to look at how R&D surveys are currently conducted and how they should be conducted to capture the country's R&D activities over the coming decade.

The CNSTAT committee has conducted its work in cooperation with a separately appointed panel of the Board on Science, Technology, and Economic Policy (STEP). The STEP panel focused on the issues of composition, structure, sourcing, and location, particularly in the context of the industrial R&D and federal funds surveys, covering the majority of U.S. funding and performance. The STEP panel's workshop, held in April 2003, informed the deliberations of this panel.

Our panel represents expertise in the fields of survey methodology, data analysis, statistics, economics, research analysis, and the organization and conduct of R&D in the public and private sectors. In selecting the members of the panel, the National Research Council sought to include representation from data users in the fields of science and technology policy and the analysis of technological change, as well as data providers from industry and academia, the two major sources of raw data. Because many of the practices and procedures for providing data depend on tax law, panel expertise in the valuation of intangible assets and the treatment of R&D in the tax code was also considered essential.

In conducting this review, the panel examined existing R&D data collection systems and relevant literature, commissioned appropriate papers, identified gaps in current methodology, and held a workshop on R&D measurement methodology. The panel held five meetings beginning in January 2003, as well as a workshop on July 24-25, 2003. The papers the panel commissioned address specific issues. In focusing on issues of statistical accuracy and reliability, the panel had the benefit of advice and analysis from two meetings with staff of NSF's Division of Science Resources Statistics and the public and private organizations directly responsible for data collection and compilation. In addition, subject-matter experts from the panel met to explore issues of statistical methodology and cognitive aspects of data collection.

The panel issued an interim report in March 2004. This interim report presented the panel's findings and conclusions regarding the present array of surveys on matters of statistical accuracy and reliability, as well as interim recommendations on near-term improvements that should be considered and could be implemented by NSF in developing plans and making resource decisions for the next several years. The interim report's major findings and recommendations have been fully incorporated into this report.

The interim report found that significant progress has been made by the Science Resources Statistics Division in fostering an environment for improvement of data quality. The panel expressed hope that these recent initiatives, buttressed by additional resources and supplemented by further initiatives such as those outlined in this report, will lay a basis for further improvements in the future.

The interim report focused on four basic methodological issues: web-based collection, the practice of providing prior-year data to survey respondents, the designation of respondents, and nonresponse adjustment and imputation.

The very newness of web-based collection in the academic and federal government surveys suggests both opportunities and challenges. Web-based collection can afford efficiencies and economies and promises to improve such functions as editing and imputation. However, in reviewing how web-based collection is now implemented, the panel raised several cautions, suggesting that additional research is needed on such issues as self-imputation that is forced by demanding a data entry in each cell.

The panel considered the practice of providing data collected in prior-year responses to respondents. Although we have some confidence that the practice does not generate significant errors, the panel urged NSF to sponsor research on the effect of imprinting prior-period data on the industrial R&D survey in conjunction with testing the introduction of web-based data collection.

The interim report found that the industrial R&D survey gives limited attention to interaction with survey respondents, beyond the largest reporters, and then only when questions or issues of nonresponse are encountered. The panel supported the initiative to identify individual respondents in companies as a first and necessary step toward developing an educational interaction with respondents, so as to improve response rates and the quality of responses. The panel also strongly recommended that NSF and the U.S. Census Bureau resume a program of field observation staff visits to a sampling of reporters to examine record-keeping practices and conduct research on how respondents fill out the forms.

The R&D surveys have very different approaches to the treatment of nonresponse and the imputation of missing values. The panel recommended that NSF revise its Statistical Guidelines for Surveys and Publications to set standards for treatment of unit nonresponse and to direct the computation of response rates for each item, prior to sample weighting.

The interim report also looked at and made recommendations in regard to each of the surveys. The major findings of the interim report are reflected in the discussions of each of these surveys in this final report.

The Panel on Research and Development Statistics at the National Science Foundation gratefully acknowledges the contributions of many individuals and organizations which supported our activities over the 2-year study period. Not all of those who gave so freely of their time, energy, and knowledge can be adequately acknowledged. As we conducted our work, it was obvious that the topic of measurement of research and development is of intense interest to a wide-ranging community of government, academic, private, and private nonprofit individuals and organizations. Without their assistance and encouragement, we could not have completed this work.

Our appreciation begins with the sponsor of the study, the Science Resources Statistics Division of the National Science Foundation. The director of this division, Lynda Carlson, was instrumental in identifying the need for the study and assisting the panel in framing the scope and approach to the study. Her insightful comments on the occasion of the first meeting of the panel presaged a very successful undertaking. She ensured that appropriate members of her staff were available to assist the panel as we approached our task, and invariably recognized the independence of the panel.

Special recognition should be given to John Jankowski, program director, Research and Development Statistics Program, who was patient, supportive, and invariably helpful in fielding the many requests for information about the programs from the panel members and staff. His key survey managers and support staff, including Rich Bennof, Leslie Christovich, Mage Machen, Ron Meeks, Francisco Moris, Brandon Schackleford, and Ray Wolfe, provided support and assistance beyond the expected. Other members of the staff of the SRS division likewise contributed their time and talents to this product. The division's assistant director, Mary Frase, as well as Fran Featherstone, Ron Fesco, Rolf Lemming, and Jeri Mulrow, assisted in laying out issues of statistical methodology and data quality. The constant interest and support of Norman Bradburn, the head of NSF's Directorate for Social, Behavioral, and Economic Sciences, was absolutely essential to the success of the enterprise. Although we recognize the crucial assistance of all the members of the NSF staff, we want to emphasize that the deliberations and recommendations of the panel are the panel's own.

Almost uniquely among federal statistical agencies, SRS relies on the services of the U.S. Census Bureau, as well as private contractors, especially QRC Macro Corporation, to perform many key functions in the surveys, including data collection. The panel, in turn, relied on the knowledgeable staff of these organizations to assist in understanding the survey operations. Staff of these organizations gave presentations at both of the panel's open meetings and participated fully in two "mini-meetings" involving members of the panel who focused on specific aspects of survey operations. The Census Bureau staff who were invariably helpful included William Bostic, Stacey Cole, Paul Hsen, Kimberly Moore, John Slanta, and Julius Smith. Adding strength and perspective to the discussion of cognitive issues in survey design at the mini-meeting on cognitive issues was Don Dillman of Washington State University.

The panel thanks QRC Macro staff members Susan Akin, Dan McMaster, Mark Morgan, and Michael Rossi, and Jim Smith of WESTAT, who freely assisted us in understanding the strengths and limitations of their operations.

The panel is also indebted to many others who made presentations on data and methodological issues, assisting us to focus on all aspects of the surveys, from concepts to uses. In addition to Lynda Carlson and John Jankowski, who participated in the initial meeting of the panel, the panel benefited from the work of Michael Bordt, Statistics Canada; Donna Fossum, RAND Corporation; Barbara Fraumeni, Bureau of Economic Analysis; Michael Gallaher, Research Triangle Institute; August Goetzfried, EUROSTAT; David Goldston, House Science Committee; Tomorhio Ijichi, University of Tokyo; Anita K. Jones, University of Virginia; Key Koizumi, American Association for the Advancement of Science; Anna Larsson, EUROSTAT; Gregory Tassey, National Institute of Science and Technology; and John Walsh, University of Tokyo.

The study director conducted many interviews with several other prominent users, providers, and producers of data on research and development. In addition to those who gave presentations at the open meetings of the panel, the panel recognizes the contributions of David Appler, Federal Laboratory Consortium for Technology Transfer; William Bonvillian, Office of Senator Lieberman; Rick Cheston, Government Accountability Office; Michael Davey, Congressional Research Service; Floyd DesChamps, Senate Committee on Commerce, Science, and Technology; Mark Herbst, Office of the Secretary of Defense; Ken LaSala, Senate Committee on Commerce, Science, and Technology; Chan Lieu, Senate Committee on Commerce, Science, and Technology; Charles Ludlam, Office of Senator Lieberman; Bob Palmer, House Committee on Science; David Radzanowski, U.S. Office of Management and Budget; Maurice Swinton, Small Business Administration; Jean Tol-Eisen, Senate Committee on Commerce, Science, and Technology; and David Trinkle, U.S. Office of Management and Budget.

The panel wishes to thank the senior staff of the General Motors Research and Development Center, which hosted members of the panel at its facilities in October 2003 for a wide-ranging discussion of the company's organization for research and development and its use of federal R&D

data. We also wish to thank Margaret Grucza, executive director of the Industrial Research Institute (IRI), for arranging for a member of the panel to discuss measurement issues at a meeting of the IRD Financial Directors Network. Our further thanks goes to Albert L. Johnson, II, chairman of the IRI Research on Research Committee, for his support of the work of the panel.

We are grateful to Barbara Bailar, who served as a consultant to the study and helped prepare portions of both the interim report and this final report. Her quality profile paper, prepared for this study, was an invaluable resource for the panel. The paper greatly assisted the panel in developing conclusions and recommendations for both the interim and final reports, and stands on its own as an important contribution to understanding the statistical characteristics of the R&D expenditure surveys. She further assisted in preparing the chapter on survey management and administrative issues. William Long and Michael McGeary also served as consultants to the panel, preparing written papers and testimony on the topics of line-of-business data and innovation surveys, respectively.

The panel owes a particular debt of gratitude to the CNSTAT staff members who supported this effort. Tom Plewes, study director, Marisa Gerstein, research assistant, and Tanya Lee, project assistant, assisted the panel in arranging six very productive meetings and producing a major interim report and this report. This team was able to assist the panel to focus on the professional literature and the issues involved with research and development statistics. Tom Plewes's experience in managing large statistical data collections was helpful in putting some of the data collection issues into context for the panel.

This report has been reviewed in draft form by individuals chosen for their diverse perspectives and technical expertise, in accordance with procedures approved by the NRC's Report Review Committee. The purpose of this independent review is to provide candid and critical comments that will assist the institution in making its report as sound as possible and to ensure that the report meets institutional standards for objectivity, evidence, and responsiveness to the study charge. The review comments and draft manuscript remain confidential to protect the integrity of the deliberative process.

We wish to thank the following individuals for their review of this report: Don A. Dillman, Social and Economic Sciences Research Center, Washington State University; Howard Garrison, Office of Public Affairs, Federation of American Societies for Experimental Biology; John Haltiwanger, Department of Economics, University of Maryland; Roderick J.A. Little, Department of Biostatistics, University of Michigan; Donald Siegel, Department of Economics, Rensselaer Polytechnic Institute; Paula Stephan, School of Policy Studies, Georgia State University; Hal S. Stern, Department of Statistics, University of California, Irvine; and Albert H.

Teich, Science and Policy Programs, American Association for the Advancement of Science.

Although the reviewers listed above have provided many constructive comments and suggestions, they were not asked to endorse the conclusions or recommendations, nor did they see the final draft of the report before its release. The review of this report was overseen by John F. Geweke, Department of Economics, University of Iowa, and R. Duncan Luce, Institute for Mathematical Behavioral Science, University of California, Irvine. Appointed by the National Research Council, they were responsible for making certain that an independent examination of this report was carried out in accordance with institutional procedures and that all review comments were carefully considered. Responsibility for the final content of the report rests entirely with the authoring committee and the institution.

Lawrence D. Brown, *Chair*
Panel on Research and
Development Statistics at the
National Science Foundation

Executive Summary

The Panel to Review Research and Development Statistics at the National Science Foundation (NSF) was asked to look at the definition of research and development (R&D), the needs and potential uses of NSF's R&D data by a variety of users, the goals of an integrated system of surveys and other data collection activities, and the quality of the data collected in the existing Science Resources Statistics (SRS) surveys. The panel has examined the portfolio of R&D expenditure surveys, identifying gaps and weaknesses, and areas of missing coverage.

The R&D expenditure surveys have owed their growth to a heightened interest in science and technology policy since World War II, as well as the growing federal involvement in R&D policy. Over the years, these data have become the accepted measures of the amounts of R&D spending, and of public and private investment in areas of science and engineering. These data have been called on to serve other purposes as well. They have become a proxy indicator of the direction of technological change. They are consulted to portray the locus of emphasis among the public, private, nonprofit, and college and university sectors. Most importantly, they are used by federal agencies, Congress, and the public to frame the national debate over the investment strategy for R&D.

SURVEY CHALLENGES

The NSF research and development expenditure data are often ill-suited for the purposes to which they have been employed. They attempt to quantify three traditional pieces of the R&D enterprise—basic research, applied research, and development—when much of the engine of innovation stems from the intersection of these components, or in the details of each. Public policy attention to early-stage technology development, the Advanced Technology Program, and process innovation requires data beyond these basic components of R&D. Similarly, the data are sometimes used to measure the *output* of R&D, when, in reality, in measuring expenditures, they reflect only one of the *inputs* to innovation and economic growth. It would be desirable to devise, test and, if possible, implement survey tools that more directly measure the economic output of R&D in terms of short-term and long-term innovation. Finally, the structure of the data collection is tied to models of R&D performance that are increasingly unrepresentative of the whole of the R&D enterprise. The growth of the service sector, the growing recognition of the role of small firms in R&D, the shift in funding from manufacturing R&D to health-related R&D, changes in geographic location, and the globalization of R&D have all served to challenge the current system for depicting the amount and character of R&D in today's economy. New forms of conducting R&D in collaborative environments, using joint ventures or outsourcing arrangements, working through alliances, and outsourcing R&D to foreign affiliates are just a few of the emerging ways of conducting research and development that are not well measured by the traditional R&D surveys.

At the same time that the foundation of R&D statistics is coming under increasing pressure, the league of uses and users continues to expand. The National Science Board continues to make sophisticated use of these data in producing the comprehensive volume, *Science and Engineering Indicators*, every 2 years, which places additional stress on the data in terms of quality and timeliness. The data are used by the administration, particularly the U.S. Office of Management and Budget (OMB) and the Office of Science and Technology Policy, to paint a complete picture of federal and nonfederal investment in R&D. Congress not only relies on the NSF data but also has directed collection of data necessary for evaluating the need for public investment in R&D. New uses of the data for purposes for which they were not originally intended are springing up. The inclusion of R&D investment in national income and product accounts, as well as in estimates of multifactor productivity, are two examples of the emerging uses that refocus attention on these data sources.

Finally, as the data have come under increasing use, they have come under increasing scrutiny. Some users are deeply troubled by the apparent discrepancy between reports of federal spending on R&D and the amounts that academia and industry report that they have received from the federal government. This large discrepancy casts doubt on the reliability of some of the data sources.

SURVEY REVIEW

Against this backdrop, the panel undertook an in-depth review of five of the recurring statistical collections by the Science Resources Statistics Division of the National Science Foundation:

- The Survey of Industrial Research and Development,
- The Survey of Federal Funds for Research and Development,
- The Survey of Federal Science and Engineering Support to Universities, Colleges, and Nonprofit Institutions,
- The Survey of Research and Development Expenditures at Universities and Colleges, and
- The Survey of Scientific and Engineering Research Facilities.

In addition, the panel considered the provisional Survey of Innovation, which has been conducted twice, in somewhat different forms.

Industry R&D Survey

The panel devoted much of its attention to the critically important survey of industrial R&D. This survey is conducted for NSF by the U.S. Census Bureau. It was last redesigned in 1991, to expand the sample into the service sector of the economy and make other changes. The panel recommends that NSF address the problems associated with this survey first, and lists its recommendations below in order of priority.

The improvements to the industry survey are extensive and expensive and call for a reconsideration of the basis for administering this survey. For several reasons, including the need to increase the professionalism of the staff of the Science Resources Statistics Division, the panel urges SRS to take the lead in the work on the industrial survey. While leaving the exact form of this more active role up to the designs of NSF and the Census Bureau, the panel suggests using the tools of the interagency agreement, the oversight of a high-quality methodological staff, and the input of highly qualified outside experts. This lead role should be undertaken while working collaboratively with the Census Bureau (**Recommendation 8.1**).

The panel strongly recommends that the National Science Foundation and the Census Bureau resume a program of field observation staff visits to a sampling of reporters to examine record-keeping practices and conduct research on how respondents fill out the forms (**Recommendation 3.11**). The first step in this process will be to make contact with respondents. The panel supports the initiative to identify individual respondents in companies as a first and necessary step toward developing an educational interaction with respondents so as to improve response rates and the quality of responses (**Recommendation 8.3**).

Although the survey has been modified and adapted over the past decade, it has largely failed to keep up with the fast-changing environment for the conduct and organization of research in the private business sector, or with advances in data collection and analysis techniques. Results from field observations should inform this redesign, and NSF should also conduct research into record-keeping practices of reporting establishments by industry and size of company to determine if they can report by more specific categories that further elaborate applied research and development, such as the categories utilized by the Department of Defense (DoD) (**Recommendation 3.1**).

NSF and the Census Bureau should test the ability to collect some disaggregated data by the newer, more detailed North American Industrial Classification System (NAICS) codes used in the industry survey today. The record-keeping practice surveys should be used to assess the feasibility and burden of providing this additional detail on industrial reporters. With this information in hand, NSF and its advisory committee should decide whether the collection of reliable R&D line-of-business data is feasible, and, if so, whether for all or a subset of reporters, and at which frequency (**Recommendation 3.5**).

The panel notes that a special emphasis panel of R&D officials in large companies, which had provided advice and spending projections during the 1980s, had been disbanded in 1990 for reasons of funding shortfalls and concern over whether the body was sufficiently representative of industrial R&D. Today, NSF has no standing advisory body to which it can turn for advice on measurement issues in the industry survey. We recommend that NSF again develop a panel of R&D experts, broadly representative of the R&D performing and R&D data-using communities, to serve as a feedback mechanism to provide advice on trends and issues of importance to maintaining the relevance of the R&D data (**Recommendation 3.8**).

Among the issues facing the managers of the industrial R&D survey is the wastefulness of surveying large numbers of establishments to find a relatively rare activity: R&D was reported only for about 3,500 of the 25,000 firms in the sample. The panel recommends the use of supplemental lists of R&D performers in drawing the sample. There are a number of practical problems to be solved in using one or more supplemental lists. Lists may overlap, and duplicates must be handled in some way. The units on the lists may not all be the same—establishments may be mixed in with companies, for example—and some editing will be needed in advance of sampling. However, the payoff in efficiency could be substantial, and the panel thinks that this approach is worth investigating (**Recommendation 3.2**). Low response rates in the survey are a concern of every user, as they may signal a problem with the quality of the estimates. The panel recommends increased reliance on mandatory reporting between economic censuses, and additional research on the topic of voluntary versus mandatory reporting (**Recommendation 8.5**).

The panel concludes that appropriate assignment of industrial classification to industrial R&D activity requires additional breakdowns of data at the business unit level. We urge NSF and the Census Bureau to evaluate the results of the initial collection of R&D data in the Company Organization Survey to determine the long-term feasibility of collecting these data.

The panel is concerned about the possibility that the editing process, replete with analyst judgment, could introduce unmeasured and undocumented errors into the publicly released data. The panel recommends that the industrial R&D editing system be redesigned so that the current problems of undocumented analyst judgment and other sources of potential error can be better understood and addressed (**Recommendation 3.12**). As NSF turns to modernizing the industrial R&D survey, the panel urges it to sponsor research into the effect of imprinting prior-period data on the industrial R&D survey in conjunction with testing the introduction of web-based data collection (**Recommendation 8.2**).

The panel took note of a recent pioneering effort to improve understanding of the impact of foreign investment in R&D in the United States by linking Census Bureau R&D data to the foreign direct investment data of the Bureau of Economic Analysis. The panel commends the three agencies for this initiative and encourages this and other opportunities to extend the usefulness of the R&D data collected by enhancing them through matching with like datasets. We urge that the data files that result from these ongoing matching operations be made available, under the protections to assure the confidentiality of individual responses that are guaranteed by the Census Bureau's Center for Economic Studies, for the conduct of individual research and analytical studies (**Recommendation 3.9**).

Innovation Survey

The panel considered the several attempts to collect data on innovation here and abroad, as well as the need for such data to illuminate the amount and outcomes of innovation activity in the economy. The panel concludes that innovation, linked activities, and outcomes can be measured and the results used to inform public debate or to support public policy development.

Furthermore, the panel recommends that resources be provided to SRS to build an internal capacity to resolve the methodological issues related to collecting innovation-related data. The panel recommends that this collection be integrated with or supplemental to the Survey of Industrial Research and Development. We also encourage SRS to work with experts in universities and public institutions who have expertise in a broad spectrum of related issues. In some cases, it may be judicious to commission case studies. In all instances, SRS is strongly encouraged to support the analysis and publication of the findings (**Recommendation 4.1**).

An additional recommendation is that SRS, within a reasonable amount of time after receiving the resources, should initiate a regular and comprehensive program of measurement and research related to innovation (**Recommendation 4.2**).

Surveys of Federal R&D Spending

In reviewing the accounting framework basis for the federal funds survey, the panel considered the growing, important uses of the federal science and technology (FS&T) budget. The panel recommends that NSF continue to collect those additional data items that are readily available in the defense agencies and expand collection of expenditures for those activities in the civilian agencies that would permit users to construct data series on FS&T expenditures in the same manner as the FS&T presentation in the president's budget documentation (**Recommendation 5.1**).

The panel reviewed the basis for collection of the data from federal agencies and compared the NSF procedures with the collection methodology employed in the RAND Research and Development in the United States (RaDiUS) database, which uses data from primary contract, grant, and cooperative agreement files as the data sources. Currently, the RaDiUS database is not adequate for obtaining estimates of federal government spending by science field. The panel urges NSF, under the auspices of the E-Government Act of 2002, to begin to work with OMB to develop guidance for standardizing the development and dissemination of R&D project data as part of an upgraded administrative records-based data system (**Recommendation 5.2**).

Similarly, the panel recommends that NSF devote attention to further researching the issues involved with converting the federal support survey into a system that aggregates microdata records taken from standardized, automated reporting systems in the key federal agencies that provide federal support to academic and nonprofit institutions (**Recommendation 5.3**).

Academic R&D Surveys

Noting that it has been some three decades since the field-of-science classification system has been updated, and that the current classification structure no longer adequately reflects the state of science and engineering fields, the panel recommends that it is now time for OMB to initiate a review of the *Classification of Fields of Science and Engineering*, last published as Directive 16 in 1978. The panel suggests that OMB appoint the Science Resources Statistics Division of NSF to serve as the lead agency for an effort that must be conducted on a government-wide basis, since the field classifications impinge on the programs of many government agencies. The fields of science should be revised after this review in a process that is mindful of the need to maintain continuity of key data series to the extent possible (**Recommendation 6.1**).

The panel recommends that NSF engage in a program of outreach to the disciplines to begin to develop a standard concept of interdisciplinary and multidisciplinary research and, on an experimental basis, initiate a program to collect this information from a subset of academic and research institutions (**Recommendation 6.2**).

We are concerned that the apparently growing collaborative environment for the conduct of R&D is not adequately reflected in the academic spending survey. The panel recommends that NSF consider the addition of periodic collection of information on industry-government-university collaborations as a supplemental inquiry to the survey of college and university R&D spending. A decision on the viability of this collection should be preceded by a program of research and testing of the collection of these data (**Recommendation 6.3**).

With regard to the academic expenditure survey, the panel observes that the exact procedure used by NSF for imputation is not well documented, but it appears that imputation is used for unit nonresponse—a practice that is highly unusual in surveys. In most surveys, unit nonresponse is handled by weighting, as it was in this survey in 1999. At a minimum, NSF is urged to compare the results of imputation and weighting procedures (**Recommendation 6.9**).

We balance our concern over the burdensome nature of the survey of academic scientific and engineering research facilities with evidence that the data have important uses, including to those who provide that data. Sensitive to these concerns, the NSF staff has recently introduced several innovations in the questionnaire and in process automation. The panel recommends that the experience in the fielding of the revised questionnaire in 2003 be carefully evaluated by outside cognitive survey design experts, and that the results of those cognitive evaluations serve as the foundation for subsequent improvements to this mandated survey (**Recommendation 6.7**). This recommendation supplements our recommendation that NSF continue to conduct a response analysis survey to determine the base quality of these new and difficult items on computer technology and cyber infrastructure, study nonresponse patterns, and make a commitment to a sustained program of research and development on these conceptual matters (**Recommendation 6.8**).

Nonprofit Sector Survey

In reviewing the attempts by NSF to collect data on the nonprofit sector, the panel noted that there were evident problems that were well documented in the methodology report on this survey. Nonetheless, the panel recommends that another attempt should be made to make a survey-based, independent estimate of the amount of R&D performed in the nonprofit sector (**Recommendation 3.10**). The panel also recommends that NSF evaluate the possibility of collecting for nonprofit institutions the same science and engineering variables that pertain to academia (**Recommendation 5.3**).

DISCREPANCY BETWEEN SURVEYS

In evaluating the potential sources of the apparent discrepancy between the federal reports of spending on R&D and the reports of performers of R&D, the panel concludes that much of the discrepancy is caused by the use of improper metrics. The panel recommends that future comparisons of federal funding and performer expenditures be based on outlays versus expenditures, not obligations versus expenditures (**Recommendation 7.1**). However, the discrepancy can be an early and important sign of problems in one or more of the surveys. The panel's recommendation is that a reconciliation of the estimates of federal outlays for R&D and performer expenditures be conducted by NSF on an annual basis (**Recommendation 7.2**).

SURVEY ADMINISTRATION AND MANAGEMENT

The panel makes several recommendations concerning the administrative and management functions of NSF with regard to the surveys. Noting that the SRS division is considered a full-fledged federal statistical agency but that it is somewhat buried in NSF, the panel nonetheless could find no compelling reason to suggest that SRS be relocated organizationally within NSF. However, we have the sense that an elevation of the visibility of the resource base for SRS would be a positive step and would serve to direct attention to the needs of the programs for sustainment and improvement.

There are several tools that NSF has in its toolbox that will help the agency gain more control over aspects of survey operations. As a start, the panel recommends that NSF, in consultation with its contractors, revise the Statistical Guidelines for Surveys and Publications to set standards for treatment of unit nonresponse and to require the computation of response rates for each item, prior to sample weighting (**Recommendation 8.4**).

The panel would like to note that significant progress has been made by the Science Resources Statistics Division in fostering an environment for the improvement of data quality. We continue to be hopeful that these recent initiatives, buttressed by additional resources and supplemented by further initiatives such as those outlined in this report, will lay a basis for further improvements in the future.

1

Introduction

This report is provided to the National Science Foundation (NSF) by the Panel on Research and Development Statistics at the National Science Foundation. The panel was convened in January 2003 by the Committee on National Statistics of the National Academies at the request of the National Science Foundation to conduct an in-depth and broad-based study of the Research and Development Statistics program of the NSF Science Resources Statistics (SRS) Division. A portion of the study was also mandated by Congress in the National Science Foundation Authorization Act of 2002.

The panel was asked to look at the definition of R&D, the needs and potential uses of R&D data by a variety of users, the goals of an integrated system of surveys and other data collection activities, and the quality of the data collected in the existing SRS surveys. Our overall review examines the portfolio of R&D expenditure surveys as a total system, finds gaps and weaknesses, and identifies areas of missing coverage.

The work of the panel is aligned with an associated workshop by the National Academies' Board on Science, Technology, and Economic Policy, which examined the current and evolving structure of the U.S. R&D enterprise.

This final report of our study contains recommendations for identifying and defining R&D activities, the appropriate goals for an integrated R&D measurement system, and recommendations on methodology, design, resources, structure, and implementation priorities.

CONGRESSIONAL MANDATE

The National Science Foundation Authorization Act of 2002 mandated that the director of NSF, in consultation with the director of the Office of Management and Budget (OMB) and the heads of other federal agencies, enter into an agreement with the National Academies to conduct a comprehensive study to determine the source of discrepancies in federal reports on obligations and actual expenditures of federal research and development funding (U.S. Congress, 2003). The legislation directed that the study examine the relevance and accuracy of reporting classifications and definitions; examine whether the classifications and definitions are used consistently across federal agencies for data gathering; and examine whether and how federal agencies use NSF funding reports, as well as any other sources of similar data used by those agencies.

In view of the fact that this committee study had been recently initiated when the legislation was passed, NSF requested and the panel accepted, the task of studying the discrepancy.

STUDY APPROACH AND SCOPE

The panel prepared an interim report, which should be considered as the basis of this final report. Indeed, highlights of the interim report's analysis and recommendations have been carried forward into this final report.

The interim report assessed the commitment of the Science Resources Statistics Division of NSF to quality of performance and professional standards and examined aspects of the statistical methodology and accuracy in the SRS portfolio of surveys. Both the interim report and this final report focus on the concept of *quality* for the NSF R&D expenditure statistics. While there is no commonly accepted definition of quality for surveys, despite over two decades of intense interest in aspects of quality in federal surveys, there is an evolving consensus that the quality of federal statistical data encompasses four components: accuracy, relevance, timeliness, and accessibility (Andersson et al., 1997).

The panel chose to focus the discussion in the interim report primarily on the dimension of *accuracy*. As defined by the OMB, accuracy includes the measurement and reporting of estimates of sampling error for sample survey programs as well as the measurement and reporting of nonsampling error, usually expressed in terms of coverage error, measurement error, nonresponse error, and processing error. The OMB working paper concludes "it is important to recognize that the accuracy of any estimate is affected by both sampling and nonsampling error" (U.S. Office of Management and Budget, 2001:1-2).

The panel adopted this perspective on the accuracy component of quality in developing its assessment of the quality of the NSF R&D surveys. We further adopted the approach to examining the sources of sampling and nonsampling error reflected in the work of Brooks and Bailar (1978). The quality profiling approach adopted by the panel was applied across the surveys, looking at common or cross-cutting issues, as well as to each of the five major current survey operations separately.

This final report also addresses the accuracy of the statistical data for each of the current and potential NSF R&D surveys but focuses more on the *relevance*, or usefulness, of the information; the *timeliness*, or the fitness for use of the data; and the *accessibility*, or utility of the statistical series, their microdata inputs, and their longitudinally matched data files. The definitions of these components of data quality are taken from the work of the Federal Committee on Statistical Methodology (U.S. Office of Management and Budget, 2001), and are used throughout this report.

In addition, this final report addresses the continuing issue of the discrepancy between the federal reports on obligations of funds for research and development and the expenditures of federal research and development funding reported by recipients of those funds. Our review of the reporting classifications and definitions used in the differing reports and a reconciliation of the data sources appear in Chapter 6.

CHANGING ENVIRONMENT FOR RESEARCH AND DEVELOPMENT

The relationship between public policy and federal statistical programs has long been symbiotic. Changing public policy changes the demand for data and the way in which data are received. Changing public policy also changes the role that data plays in the formulation of public policy. In turn, statistical programs help identify the need for and assist in assessing the requirements and results of public policy. Data help illuminate the paths that public policy should take to solve public issues.

Over the past half century, the federal government's statistical programs that produce data on research and development have exemplified this relationship. As interest in the federal role in science and engineering has mounted and focus on the contribution of research and development to innovation and national growth has intensified, the need for good data to measure the research and development enterprise has grown (National Research Council, 2000). As data have become more available and the sophistication of means of disseminating the results of the data collections has expanded, the public's interest in issues of trend, level, structure, balance, equity, and appropriateness of the public and private investment in research and development have intensified. To understand the current status of research and development statistics at the National Science Foundation, it is important to understand the context in which the statistical programs have developed, and the role they play in contributing to the understanding of public policy issues today.

World War II is often identified as a watershed in U.S. science policy. Brooks (1968) describes federal policy toward science in the prewar period as largely instrumental in character—concerned with utilization of science and technology for closely defined purposes, such as agriculture, defense, and natural resource development. Although there were examples of federal investment in basic science, such as the National Bureau of Standards, which was doing critical basic research in a broad spectrum of areas, the sense of mutual dependence with the scientific community had not yet fully blossomed.

The tenor of the times, as U.S. science and engineering policy stood on the springboard that would catapult the enterprise to new levels, was best captured in *Science—the Endless Frontier*, a report by Vannevar Bush (1945), who had headed the wartime research and development programs. The influence of this study, which was instrumental in the establishment of the National Science Foundation 5 years later, cannot be overestimated (National Research Council, 1995). Importantly, the study illuminated both the times and the level of understanding about the issue of research and development.

The data on research and development were hardly up to the task at hand when Bush conceptualized the future organizational structure for federal support of research and development in 1945. For example, information in his report about trends in expenditures in industry, academic institutions, and the government were the "best estimates available," and they were "taken from a variety of sources and arbitrary definitions have necessarily been applied" (Bush, 1945:20). In order to obtain factual information concerning research expenditures in colleges and universities for the report, questionnaires were sent to a list of colleges and universities accredited by the Association of American Universities. Some 188 replied, and 125 reported organized research programs (Bush, 1945). At the time of this report, federal expenditures for research and development had been compiled from budget sources in a fairly consistent manner for two decades, although there were recognized shortfalls in the data and omissions in that they did not include grants to educational institutions and schools (Bush, 1945).

The years since publication of the Bush study have been marked by several major thrusts that have had a remarkable influence on the course of research and development in the United States, and, in turn, on the requirements for measurement of the research and development enterprise. This was a period, characterized by the Cold War and international competition, in which the federal government funded an increasing share of research in the nation's universities. This "second mega-era" for science and science funding was further characterized by two other forces—the virtual explosion of federal support for science and engineering, and the birth of the National Science Foundation—that had an especially pronounced effect on the demand for and eventual availability of information on research and development in the United States (U.S. Congress, House, 1998:9).

Explosion of Federal Support for R&D

The growth of federal support was sure yet uneven in scope and impact in the decades since the Bush study. In the immediate aftermath of the war, even before the birth of the National Science Foundation in 1950, the trend toward consolidation of federal research in a few agencies became apparent. The National Institutes of Health established prominence in health-related research, including university-based biomedical research and training; the Office of Naval Research took on a major role in supporting academic research in the physical sciences; and the Atomic Energy Commission took on most of the research in atomic weapons and nuclear power (National Research Council, 1995). The National Aeronautics and Space Administration and the Advanced Research Projects Agency in the Department of Defense were signs of further consolidation and growth following the launch of *Sputnik* by the Soviet Union in 1957, which "provoked national anxiety about a loss of U.S. technical superiority and led to immediate efforts to expand U.S. R&D, science and engineering, and technology deployment" (National Research Council, 1995:42).

Each of the decades that followed had a theme and an emphasis that further challenged the data systems that helped to illuminate the scope and trends in R&D. A renewed emphasis on health and biomedical research in the 1950s, an increased focus on environmental and energy issues in the 1970s, a new national commitment to international competitiveness and breakthroughs in information technology in the 1980s and 1990s, and then the robust growth of the health and medical research sector since the 1980s were examples of the global shifts in emphasis and resourcing that needed to be explored and explained.

During this same time period, the increasing recognition of the contribution of R&D to economic growth and productivity advanced heightened national interest in R&D and, in particular, innovation. Over time, the importance of information on R&D expenditures was recognized as an indicator of the extent to which the generation and diffusion of knowledge has become an economic activity with the growth in the science and engineering enterprise (Rosenberg, 1994).

A fundamental importance of R&D in the process of economic growth is that it generates spillover benefits. As documented by Griliches (1980) and Mansfield (1980), the benefits of innovation made by one company can spread widely, or spill over, to many industries. For example, cost-reducing production techniques developed in one industry are likely to be copied or adapted for use by companies in many industries (U.S. Congress, Joint Economic Committee, 1999a). The existence of spillover benefits explains why studies find that investment returns from R&D to the economy as a whole are often greater than the returns to the investing businesses themselves.

In order to estimate spillover effects, two types of benefits from innovation must be identified: those flowing from "customer" benefits and "knowledge" spillovers. Customer spillovers occur when customers benefit from new products and better production techniques that businesses develop in order to earn higher rates of return. The value of the benefit is greater than the cost to the customer. On a global scale, these customer benefits have been embodied in imported capital and intermediate goods, contributing significantly to economic growth.

The rapid exchange of technical information facilitated by academic journals, conferences, the Internet, and licensing agreements has generated knowledge spillovers, which help firms increase productivity more broadly, even with patent protections for the innovating company.

Growing Importance of R&D Expenditure Measures

Data on R&D expenditures have documented these trends in the size and scope of science and engineering, as well as the contribution of R&D to growth. As federal government support for science and engineering grew, so did demand for information to support decision making and the burden on the instruments for measuring science and engineering activity.

The birth of the National Science Foundation in 1950 was quickly followed by an articulated demand for measuring the science and engineering enterprise in the United States. The first emphasis was on identifying the federal contributions in a systematic way, so NSF quickly assumed responsibility for maintaining the National Register of Scientific and Engineering Personnel and commissioned the first collection of data on federal R&D funding and R&D performance. The Survey of Federal Funds for Research and Development, which collects data on R&D obligations made by federal agencies, was instituted in 1953. In that same year, NSF set a pattern for future statistical operations by contracting out the first Survey of Industrial Research and Development to the Bureau of Labor Statistics. Administration of the survey was later transferred to the U.S. Census Bureau. In the same year, NSF conducted the first of six occasional surveys of R&D performance by nonprofit institutions. In 1954, NSF conducted the first small-scale surveys of R&D at major universities.

Looking back on the 1950s, it is apparent that a pent-up demand for information on science and engineering led to an era of statistical innovation, enabled by the fresh infusion of resources from the new NSF. It was a period of creativity and risk-taking as NSF expanded its portfolios in both the expenditures and human resources statistics programs. This era saw the birth of statistical data collections that still, to this day, define the base of knowledge about science and engineering expenditures. And it established precedent in the organizational structure and operational modes of NSF that continue to define NSF's approach to survey operations.

Knowledge begets the demand for more knowledge. Over the next two decades, NSF expanded the data it collected on public support for science and engineering. A second predominant initiative was to deepen the data collected on federal R&D spending by expanding detailed fields of science and engineering as well as budget function, as well as by collecting federal obligations for research to universities and colleges by agency and detailed field.

In the mid-1960s Congress initiated a role that NSF has continued to play with regard to R&D statistics when it mandated the Survey of Federal Science and Engineering Support to Universities and Colleges. The purpose of this data collection was to enable Congress to better understand the role of the federal government in supporting academic research and development. Congress continued to mandate the collection of additional data on R&D performance and infrastructure in the 1980s and early 1990s: the National Survey of Academic Research Instruments and Instrumentation Needs in 1983; what is now the Survey of Science and Engineering Research Facilities at Colleges and Universities in 1986; and the Master List of Federally Funded Research and Development Centers in 1990. In many ways, congressional interest has shaped the landscape for research and development measurement. Congress has played an activist role in formulating NSF's R&D expenditure survey portfolio by supporting the collection of specific data and by relying on the resulting data in its formulation of policy.

Simultaneously, NSF added several collections of human resources data. The Survey of Earned Doctorates was begun in 1957; the annual Survey of Graduate Students in Science and Engineering began in 1966; and the Postcensal Survey of Scientists and Engineers commenced in 1982. It was replaced by the umbrella Scientists and Engineers Statistical Data System (SESTAT) in 1993, which incorporated data from the National Survey of College Graduates, the Survey of Recent College Graduates, and the Survey of Doctorate Recipients. The quest for improvement of the surveys on the human resources side parallels this study. The Committee on National Statistics recently completed a study of options for improving the SESTAT database, recommending a best approach to carry the database through the next decade and encouraging NSF to pursue opportunities to improve understanding of the numbers and characteristics of scientists and engineers in the United States (National Research Council, 2003).

Several innovative collection efforts were mounted by NSF over the years. Some were absorbed into or merged with regular data collections; some simply fell by the wayside. A survey of Science and Engineering Activities at Universities and Colleges was appended to the Survey of Industrial Research and Development in 1964 and absorbed into a new annual Survey of Research and Development Expenditures at Universities and Colleges in 1972. In 1971, SRS started an Industrial Panel on Science and Technology with about 80 members to obtain quick, qualitative information on issues in industrial R&D (National Research Council, 2000). This panel was disbanded in 1991.

Nearly a half a century of growing demand for understanding the impact of the science and engineering enterprise on the economy shaped today's portfolio of research and development expenditure surveys. The growth has largely been responsive to specific congressional or agency needs, as each new data collection activity was initiated to address a narrow topic. In its 2000 report, the National Research Council's Committee to Assess the Portfolio of the Division of Science Resources Studies of the NSF observed that this evolution yielded several rather unconnected data collections in which surveys failed to "serve as a piece of a cohesive R&D funding and performance data system" (National Research Council, 2000:25). As an example, the committee discussed the discrepancies in funding and performance estimates among its surveys, which are covered in some depth in Chapter 6 of this report. The committee concluded that "further work to improve comparability—even the integration—of these surveys would improve their analytic value" (National Research Council, 2000:25).

In retrospect it is apparent that R&D expenditure statistics progressed in tandem with the growing federal involvement in R&D. They soon became the accepted measures of the amounts of R&D spending and public and private investment in areas of science and engineering. These data were called on to serve other purposes as well. They became a proxy indicator of the direction of technological change. They portrayed the locus of emphasis among the public, private, nonprofit, and college and university sectors. Most importantly, they helped frame the national debate over the investment strategy for R&D.

There are other reasons for the growing importance of R&D expenditure measures in recent years. The science and engineering enterprise has changed substantially in the past two decades, and there is an evident association of trends in research and development spending with questions of national growth and prosperity. At the same time, the task of measuring R&D activity has become more diffuse and complicated. For example, the declining share of R&D expenditures contributed by the federal government has meant that federal agencies no longer account for the bulk of R&D spending. Other major changes often cited in the literature are the shift from a manufacturing to a service economy, the diffusion of innovation among smaller firms, the shift from emphasis on manufacturing and engineering R&D investment to medical sciences, the geographical clustering of domestic technological research and innovation—the so-called Silicon Valley effect—as well as the globalization of R&D activity and innovation. These are indeed challenging times for those in the National Science Foundation and elsewhere who are responsible for the provision of data to illuminate these changes as they occur.

Declining Federal Share in R&D Spending

In 1979, the federal government, which had been the main investor in research and development since the expansion of funding in the post-World War II period, saw its proportion of spending on R&D fall below 50 percent. The federal government had accounted for about two-thirds of R&D spending as recently as 1968, but since then the share has fairly steadily declined while industry's share has steadily increased. In 2002, the latest data available, the federal government's total spending accounted for just 26 percent of total U.S. R&D expenditures (see Figure 1-1).

The strong growth of industry-funded R&D since the early 1980s may be attributable to a number of factors (U.S. Congress, Joint Economic Committee, 1999a). Among the factors that have been named are lower regulatory costs and higher competition in many U.S. industries, increasing global integration and greater global competition, and the fast-changing technological nature of today's manufacturing processes and markets. The result of this shift to industry funding is a greater need for understanding such aspects of R&D expenditures as the relationship between R&D spending and the general state of the economy, the trends by sector and size of establishment, and the specific geographical impact of these trends.

This fact of diminishing importance of the federal government in the R&D enterprise has several practical implications for data collection activities. Most obviously, in contrast to just three decades ago, it is no longer possible to simply collect data that are administratively available from the 32 federal agencies that conduct most of the government-funded research and development programs and be assured that the majority of U.S. R&D activity is being tallied.

The growth of influence of the private sector has compounded the difficulty of accounting for R&D expenditures. Industrial funding is more diffuse and less easily measured. Although R&D activity in industry is concentrated in a few hundred large companies, an inclusive tally of indus-

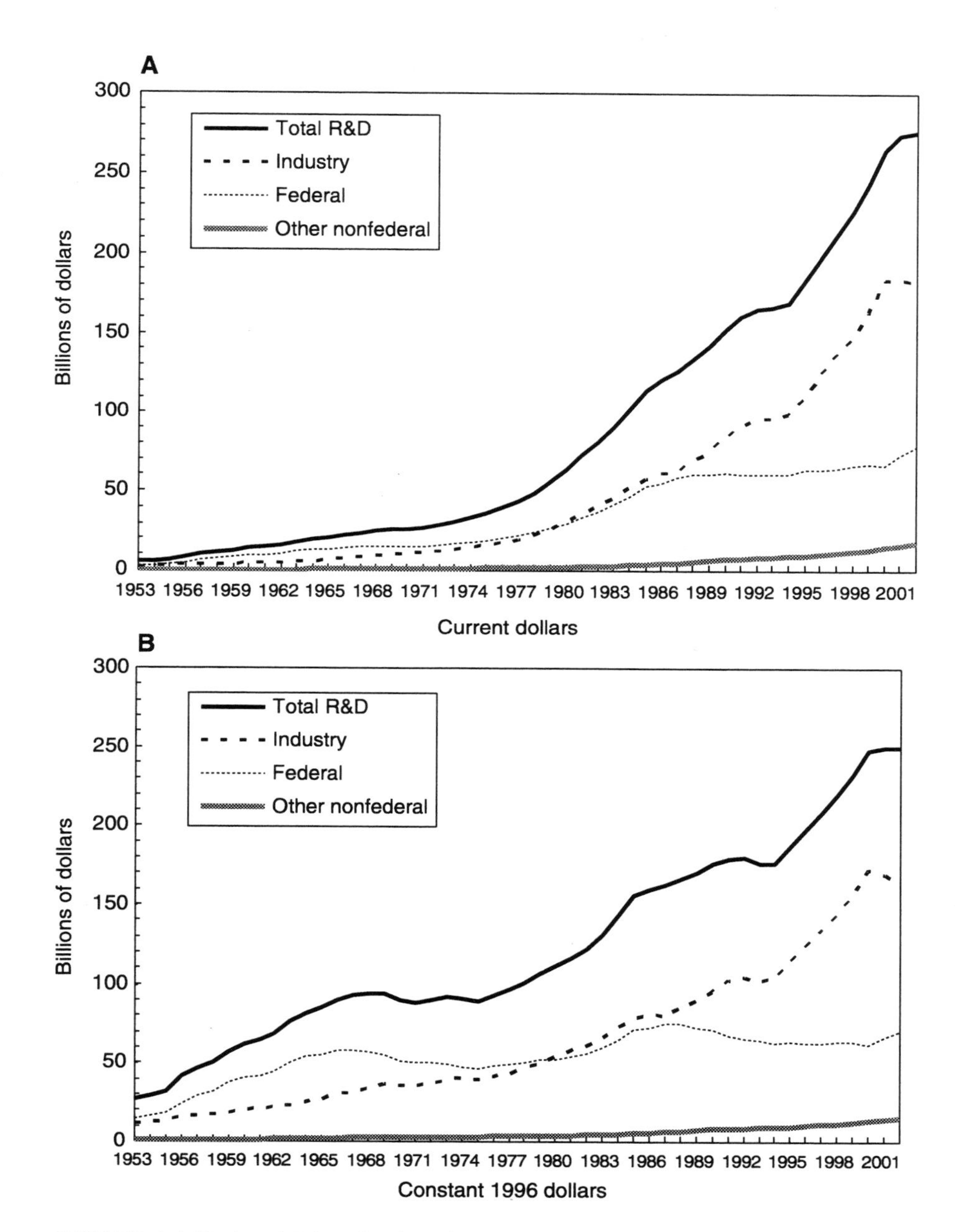

FIGURE 1-1 National R&D funding, by source: 1953-2002.
SOURCE: National Science Board, 2004: Figure 4-3.

trial R&D activity requires that surveys cover many sectors of activity and thousands of smaller establishments, some in their infancy.[1] Until 1990, in fact, the Survey of Research and Development in Industry collected data from only a small number of large firms. To expand coverage of the universe of companies that conduct R&D in the private sector, the survey was redesigned in 1991. Within the increasingly important private sector, fast-moving changes in structure continue to challenge accurate and informative data collections.

Shift to a Service Economy

The structural shift from a goods-producing to a service-producing economy has been a powerful force in shaping U.S. institutions for several decades and has been amply documented (National Science Foundation, 2001). The service sector of the economy has eclipsed the goods-producing sector in proportion of gross domestic product (GDP), now accounting for over 77 percent of it (Jankowski, 2002). This growth of economic activity outside the classic areas of agriculture, mining, construction, and manufacturing is reflected in the growing contribution of the service sector to R&D activity. According to current definitions and data, nonmanufacturing business R&D now accounts for about 39 percent of R&D performance, up significantly from a 5 percent share in 1983. Although this trend has been most evident in the United States, there is evidence that it is widespread. The Organisation for Economic Co-operation and Development, using its definitions, has reported that the service sector accounted for 34 percent of industrial R&D in the United States in 2000; the figures were comparable for Canada (28 percent) and Australia (27 percent); while the proportions in Great Britain (17 percent), Germany (5 percent), and Japan (3 percent) were somewhat less.

This increasing presence of a dynamic service sector in the economy has had ramifications for many aspects of economic measurement, from necessitating design changes in the economic censuses, to refocusing GDP and employment measurement, to the development of the North American Industrial Classification System (NAICS) in 1997 to replace the old Standard Industrial Classification (SIC) system. The measurement of R&D should be no different in this respect. For many years, the emphasis of R&D data collection and analysis had been devoted to measuring R&D in the manufacturing sector, which was seen as the source of most innovation and, by the way, produced tangible outputs which were easier to measure than

[1] It is estimated that the largest 100 companies account for 54 percent of R&D spending.

intangible services. This focus of attention resulted in a view of the service sector as lagging in technological innovation and consequently in productivity growth (Tether and Metcalfe, 2002). Thus, the relative difficulty of measuring R&D in the service sector has fostered a view that was at variance with the anecdotal observation that the service sector is a very large consumer of new technologies and human capital and innovative activity. The service sector not only exhibits considerable innovation in such activities as the Internet, web-based services, and new financial instruments, but also appears to be providing innovative services to client industries, such as systems integration services in manufacturing.

The growth of the service sector not only presents issues of classification and measurement, but also brings into question the applicability of the traditional definition of R&D. Jankowski points out that the service sector creates value and competes by buying products and assembling them into a system or network, efficiently running or operating the system, and providing services for customers who are often members of the general public. Thus, the R&D portfolios of service firms reflect these required core competencies and typically address: system design (network of physical products and information); system operation (equivalent to manufacturing processes in manufacturing industries); and service design and delivery, including interaction with individual customers. It is an open question whether many of these service activities would be interpreted to be readily classified as traditionally measured R&D in surveys.

The growth of the service sector confounds the identification of the share of innovative activity that is accounted for by traditionally defined R&D. Many of the innovative inputs of the service sector are not currently classified as R&D. These include market research, training in innovation, adoption and adaptation of new technology, start-up activities, organizational changes, and incremental impacts. There are, however, major definitions and measurement issues here. For example, Gallaher and Petrusa (2003) suggest that customization, an important product of the service sector, is a gray area for distinguishing between R&D and non-R&D innovation activities.

By nature, R&D in the service sector is more difficult to measure. Michael Gallaher, in a presentation to the panel, outlined several of the sources of difficulty in presenting the initial results of his research into service sector measurement issues. It focused on four industries: telecommunications (NAICS 5133), financial services (NAICS 52 and 53), computer systems design and related services (NAICS 5415), and scientific research and development services (NAICS 5417). His research identified issues of misclassification and mismeasurement that pertained to each of these industries (Table 1-1).

Greater recognition of differences between the growing service indus-

TABLE 1-1 Measurement Issues in Four Service Industries

Industry	Misclassification Issues	Measurement Issues
Telecommunications	Content developers	Larger companies with R&D division
Financial services	Patent holding companies	R&D closely linked to new product offerings
Computer systems design and related services	Computer hardware, computer software (shrink wrap)	R&D is frequently "on the job" integrated with the service
Scientific research and development services	R&D outsourcing, biotech startups	High percentage of activities are R&D

SOURCE: Gallaher, 2003.

tries and the more traditional manufacturing sector may lead to the recognition that different concepts and definitions as well as techniques and methodologies are needed to measure R&D expenditures in the service sector. With further research of the type discussed above, NSF may come to recognize the need to consider collecting information from the service sector using a very different survey from the current collection instruments.

ISSUES IN THE FINANCIAL SERVICES SECTOR

Some of the most key assets in financial services have been new ideas. New ideas produced derivative products, securitizations, mutual funds, exchange traded funds, and thousands of new securities. The power of ideas has not only made financial service firms well-off, but also enabled savers to more efficiently invest, firms to raise capital, and homeowners to enjoy lower mortgage rates.

Patents are one traditional method of defining and measuring innovation. But since the early days of the 20th century, there has been considerable doubt in the United States as to whether methods of doing business fell under the definition of patentable subject matter, namely "any new and useful process, machine, manufacture, or composition of matter." While the United States did not explicitly forbid so-called business method patents, as many nations in the French and German legal traditions did, there was still a presumption that they did not fall into these four categories and hence were not patentable. To make matters worse, financial service firms were required by the marketplace and by regulation to make public the broad outlines of publicly registered securities through prospectus disclosure.

As a result, financial innovators took different approaches to managing their key intellectual assets than did other industries. It was well understood that ideas alone were not likely to generate competitive advantages. Key innovations by one investment bank were rapidly copied by their competitors, with imitative products usually occurring soon after the innovator had an opportunity to bring to market a single issue of a new security. Firms boasted that they could use the information in required securities filings to "reverse engineer" virtually any discovery. Whole prospectuses were copied virtually verbatim. Firms used branding, in the form of service-marked products, to differentiate their offerings from one another. Defecting employees would frequent carry knowledge with them from one employer to the next, while vendors would transfer insights developed during projects with financial institutions to their competitors. While firms tried to protect insights (e.g., trading strategies) through the use of trade secrets, these protections were frequently modest at best. While most have described the latter 20th century as a time of great financial innovation, other dissenting voices have countered that quite little *revolutionary* innovation took place in financial markets, in part because there were few incentives for firms to invest massive amounts of money in doing R&D that could be easily appropriated by rivals.

There are also, it should be noted, considerable differences among rivals. Some firms pride themselves on creating new products and processes; others excel at scanning the markets and then finding creative ways to reverse engineer others' products, rebrand them, and push them through their powerful distribution networks.

Unlike traditional manufacturers, for which most innovations grow out of dedicated R&D laboratories, more of the innovation in financial institutions comes from within operating units. With a centralized R&D function, it is fairly easy to keep track of the inventory of intellectual property and R&D. The challenge for data collection within financial service firms is, however, much larger: they often have no clear sense of the amount of new product and service development activity going on, or even the number of successful activities completed.

Today, these time-honored arrangements are under stress. Since the *State Street* decision legalizing financial patents in 1998, hundreds of financial patents have been issued and many more applications are pending (*State Street Bank & Trust Co. v. Signature Financial Group, Inc.*, 927 F. Supp. 502, 38 USPQ2d 1530). These awards are proving to have material economic value. The case that ignited this new change in policy pitted State Street Bank against a smaller firm, Signature Financial. When Signature won its suit to stop State Street from infringing Signature's patent on a way of creating mutual funds, State Street's stock fell 2.3 percent or $277 million in one day. More recently, the trading firm Cantor Fitzgerald received

$30 million from the Chicago Board of Trade and the Chicago Mercantile Exchange to settle charges of patent infringement. To date, small and large firms, and even academic institutions, have been rushing to patent their financial products and processes like homesteaders on a frontier of financial intellectual property. How this will affect the innovation process remains unclear, but it is likely to be important. This is an example of a sectoral picture that NSF needs to draw to fully explain trends in R&D and identify emerging measurement issues.

Role of Small Firms in R&D

Another significant change in the environment for R&D, closely associated with the growth of the service sector, has been the growth in the role of small firms in the generation of innovation and economic growth through R&D. Firms in the service sector that perform R&D tend to have fewer employees and smaller R&D budgets than those in the manufacturing sector. Some 43 percent of currently measured R&D in the service sector was performed by firms with fewer than 500 employees, compared with a contribution of about 8 percent by smaller firms in the manufacturing sector.

The important question for purposes of economic measurement hinges on the role of firm size, or market concentration, on R&D investment and innovative performance. Is large size conducive to R&D investment, as hypothesized by Schumpeter (1947)? Although large firms account for the preponderance of R&D in the economy, a literature review by Achs and Audretsch (2003) found that empirical evidence suggests that small entrepreneurial firms play a key role in generating innovations in selected industries.

More importantly, the new recognition of the role of small businesses has been postulated to result from a class of studies that challenge traditional R&D measurement. The main instruments for measuring R&D activity are restricted to measuring *inputs* into the innovative process, such as R&D expenditures. Achs and Audretsch (2003) suggest that it was only when measures were developed to measure innovative *output* that the vital contribution of small firms became recognized. For example, research on patent activity by smaller firms suggests the presence of an important class of "serial innovators" whose contribution to technical progress is both effective and important within their industries. It is claimed that this contribution is not adequately measured through reporting of R&D expenditures (CHI Research, Inc., 2002). In this view, activity in small firms is better depicted through measures of innovation (see Chapter 4).

Increasing Role of Medical R&D

No trend in federal funding of research and development has been as pervasive as the shift in the spending for health-related R&D. Over the past two decades, health-related R&D rose from about one-quarter (27.5 percent in 1982) to nearly one-half (45.6 percent in 2001) of the federal R&D budget. In order to fully understand the scope and impact of this funding shift, it is important to understand the "what" (the reason for this shift) and the where (which fields, performers, and geographic locations have benefited).

Medical research and development has received the lion's share of increases in federal support (National Science Board, 2002). The federal decision to grow the budget for medical R&D was related to successive campaigns to eradicate cancer, then AIDS, then other diseases over the past several decades. Biotechnological advances, the human genome project and its offshoots, and, most recently, investments in activities related to countering bioterrorism have also played a role (National Science Board, 2002).

Health-related R&D is skewed toward basic research, which accounted for over 55 percent of the budget authority for health in fiscal year (FY) 2001. Not surprisingly, the majority of federal health-related support is directed toward academia. The result of these shifts in funding is that, in FY 2001, most (61.9 percent) of federal support to academia was funneled through the U.S. Department of Health and Human Services. At a time when funding for most disciplines has been stagnant, in real terms, obligations for the life sciences have nearly doubled since the early 1990s. There is clear evidence that this infusion of funds has, in turn, attracted a larger proportion of graduate researchers.

This trend has a significant impact on measurement. In order to understand the effect of this shift, timely information is needed not only on federal spending by budget authority, activity, agency, performer, and field, but also on the associated science and engineering human resource investment.

It is difficult to account for the impact of the increasing investment in health-related R&D in the public sector because of a lack of a common sectoral definition. It is almost impossible to identify trends with precision in the business sector because there is no accepted definition of a "biomedical industry," which would certainly contain bits and pieces of health services, pharmaceuticals, and other industries classified elsewhere by NAICS code. In 1999, companies officially categorized in the "health care services" industry accounted for only 0.4 percent of all industrial R&D. It is difficult to appropriately portray a business R&D counterpart to the public R&D investment, in part, because pharmaceuticals and medicine are classified as manufacturing in the chemical industries, and, in part, because the NAICS categories are not sufficiently detailed to identify pieces of the sector.

There are other examples of industrial sectors that are similarly hard to document—one being software development, which is conducted in various ways in a broad spectrum of industries. It has been suggested that these cross-sectoral portraits could be obtained by collection of more detail on line of business within the NAICS industries and collection of information on the fields of science from industry, as are the data for the federal government and academia. The issues attending line-of-business data collection are discussed in Chapter 3. There are several reasons that data on field of science in industry have not been collected. Suffice it to say that, until such data are available for the private as well as the public sector, understanding of the impact of emerging sectoral trends will always be incomplete.

Geographic Clustering

The emergence of the phenomena of Silicon Valley, the Route 128 corridor, and several other centers of high technology, often concentrated in the vicinity of major research-oriented academic institutions, has contributed to interest in the geographic aspects of R&D. The reasons for this interest are varied. Local economic development interests seek to attract and build these concentrations for their economic growth and job-producing potential. Policy makers at all levels seek to understand the workings of clusters so as to better direct public investment and land use policies. And businesses seek to exploit clustering for the several benefits these clusters afford: the ability to quickly capture spillovers from like industries; the availability of venture capital sources or "bands of angels"; proximity of incubator firms; and the prospect of parenting of new firms by larger, successful organizations (Thornton and Flynn, 2003:5).

The clusters are variously defined by region, state, and locality (Krugman, 2000). They are often defined by proximity to research universities (Acs et al., 2002). They may be unplanned or planned, like some science parks in the United States. The fact of clustering and the intense interest in the extent and determinates of clustering put a premium on understanding patterns of R&D expenditure by geographic area.

In an attempt to identify the geographic location of R&D activity, NSF has developed and published data geographically separated by state on R&D performance by industry, academia, and federal agencies, along with federally funded R&D activities of nonprofit institutions. These data show that R&D is substantially concentrated in a small number of states. NSF also prepares and publishes two analytical measures—state R&D level as a proportion of gross state product and the proportion of R&D by sector—in order to depict differences in "intensity" and source of R&D. Although the state-level data may be useful for comparison purposes, they contain measurement error. Even if reliable, they cannot be adequate to depict the extent and impact of geographic clustering, which occurs at the local level and sometimes crosses state lines.

Two recent studies on the geographic distribution of R&D activities illustrate why it is informative to disaggregate R&D by locality. In a study for the U.S. Economic Development Administration, Reamer et al. (2003)

examined the impact of R&D activity in local areas. They found that almost all innovation takes place in metropolitan areas, and that large metropolitan areas have an advantage in the innovation process due to their greater specialization and diversity. The presence of a public R&D institution (university, nonprofit research institute, federal laboratory) did not necessarily lead to local corporate technology spillover effects in all areas. Rural and small metropolitan areas did not benefit as much from the presence of such research institutions as did larger metropolitan areas.

Likewise, a pioneering compilation of detailed federal R&D activities (laboratories, centers, universities, and companies performing federally funded R&D) by local area used the RAND Research and Development in the United States (RaDiUS) database and supplemental data sources to detail the geographic aspects of federal R&D investments (Fossum et al., 2000). The report portrayed an impact of federal R&D funding that was heavily concentrated in a few regions, in spite of being spread in some way into nearly every community in the nation. Despite these recent attempts to put a geographic face on R&D activity, the data by state and substate level are not considered sufficiently robust or reliable to support investigation of the extent and impact of geographic clustering.

New Organizational Structures for R&D

The very vibrancy of the R&D enterprise owes significantly to the vibrancy of business, social, and political arrangements that initiate, support, integrate, and sustain innovative activity. These arrangements are constantly evolving, much as living organisms adapt and change to survive and prosper. Organizations learn and evolve. The successful adaptors join, split, move, and change associations quickly and efficiently. While this is natural and necessary, it does make life quite difficult for those who attempt to measure various economic phenomena.

The impact of alternative and emerging organizational structures is potentially a more difficult issue for the measurement of R&D than for many other types of economic activity. The organizations that conduct intangible activities like research are sometimes easily located in corporations and other entities with accounting systems that permit independent reporting of those activities. An example of an organization that is generally identifiable and for which data exist independently is the research laboratory.

Quite often, however, the organizations that conduct development activities are deeply embedded in organizations that are not traditionally considered wellsprings of R&D activity, such as manufacturing operations and marketing. As a result, these development activities are not easily distinguished from production and other corporate processes. They are often overlooked because they are not easily identified in the accounting systems of most companies, which do not usually permit separate identification of R&D expenditures and benefits.

Other research activities cross traditional organizational lines and are similarly difficult to identify and describe. Examples of this growing group of R&D organizational entities are those formed by cooperative agreements, domestic and international collaborations, alliances, and strategic research partnerships. The entities formed by these arrangements permit the pooling of assets and the sharing of knowledge irregardless of organizational lines and geographic boundaries.

In his recent work, Chesbrough terms these arrangements "open innovation" (Chesbrough, 2003). In contrast with the closed innovation stereotype of an embedded corporate research center that seeks to single-handedly develop, commercialize, and dominate an emerging technology, the open innovation companies combine their internal capabilities with an awareness of the innovation marketplace and a business model that embraces licensing, acquisitions, and collaboration to maximize the speed and impact of innovation. New forms of innovation may allow companies to reduce spending on internal R&D facilities while continuing active R&D programs.

When carried into other areas of the business, at their ultimate, these arrangements define a new kind of 21st century enterprise that separates its physical and legal boundaries and fragments the old legal entity into a labyrinth of licenses, contract, and other trading agreements, often involving multiple jurisdictions (Eustace, 2003). These trends are further characterized by Romer as being part of a "soft revolution," in which assets described as soft, immaterial, or intangible may be key economic goods (Romer, 2003). An analysis by the European Community suggests that this "hidden" productive economy that is intimately related to R&D requires new measurement tools. Its report found that these new arrangements would not be appropriately depicted in current measures.

Today's R&D expenditure surveys produce measures based on firms that conduct R&D in traditional entities, such as labs, and are operated within fixed boundaries with resources that are physical and owned. To the extent that they fail to identify the new organizational structures, they give a flawed picture of innovative activity. Estimates of total innovative activity by Hollander and Mansfield suggest that focusing on "dedicated" R&D effort may be missing half of all innovative activity.

Globalization of R&D Activity

One of the most important and complex trends is the so-called globalization of R&D and innovation. There is much testimony to an apparent rise in cross-border corporate R&D, although past studies have largely been anecdotal or based on very small samples. In a paper presented to the National Research Council workshop on R&D data issues, Kuemmerle (2003) summarized the results of a study of the foreign direct investment decisions of 32 multinational firms. He found extensive globalization of activity among his sample of multinational corporations, with an average of nearly five international R&D sites per firm. These firms invested in R&D activities abroad either to appropriate relevant knowledge and skills through externalities, or to transfer knowledge and skills within the company from knowledge creation to manufacturing and marketing. His report concluded with a recommendation for more detailed data collection and the standardization of data collection on foreign direct investment in R&D across more countries.

There are several indirect indicators on the extent of globalization of R&D, primarily through three major data sources: counts of international business alliances maintained in the Cooperative Agreements and Technology Indicators database compiled by the Maastricht Economic Research Institute on Innovation and Technology (CATI-MERIT); records of government-to-government cooperation maintained by RAND; and data from NSF on U.S. expenditures financed by foreign sources, industrial R&D performed by foreign affiliates of U.S. companies, and industrial R&D performed by U.S. affiliates of foreign companies by NSF. These data sources depict a large amount of international activity but by no means an explosion in the internationalization of R&D. In fact, these indirect measures suggest a leveling off in new activity following a peak in the mid-1990s. The importance of the international sector suggests a need for more and better expenditure measures that cross international borders.

There is no doubt that the globalization of R&D will have a significant impact on the measurement of R&D. A fresh examination of the impact of globalization by the Conference Board, which is reviewing the validity of science and technology indicators for international comparisons and for conducting R&D surveys of multinational firms, has concluded that substantial organizational change in multinationals is affecting the way R&D is conducted (McGuckin, 2004). The study observes that research is becoming more closely aligned with corporate business plans, and that basic research is not very relevant to most firms since multinational businesses rarely undertake research that does not have some potential for application. Research is increasingly being conducted in closer collaboration with outsiders (universities and business alliances), and development makes extensive use of alliances, particularly customer and supplier partnerships. Some of the costs of R&D are being borne by suppliers and probably not identified as R&D on their books.

Some of the concern focuses on the global outsourcing and offshoring of various aspects of U.S. business operations, including R&D. It has been suggested that research and technology outsourcing is coming of age, and a great deal has been written about the subject of external sourcing of R&D (Howells, 1999). However, little is known about the extent of global outsourcing of R&D. There is an expanding body of knowledge—some coming from the Sloan Foundation Industry Centers program—that suggests that outsourcing of R&D is becoming an issue of some importance. This path of inquiry suggests the need for the systematic collection of information on this phenomenon.

CONCEPTS AND DEFINITIONS

Understanding of the changing environment for research and development is conditioned by the concepts and definitions that underscore the measurement of R&D. These concepts and definitions have been shaped, in turn, by the changing environment for research and development.

This symbiotic relationship between the phenomenon to be measured and the measures themselves constitutes a continuing challenge to NSF and the science and technology constituents the agency serves. On one hand, it is important that concepts reflect changing reality, and definitions are revised as conditions change and understanding advances. On the other, it is important that concepts and definitions are consistent with those used in other countries, particularly U.S. trading partners; that they are consistent over time so that intertemporal comparisons can be made; and that they represent a general professional and public understanding of the phenomena to be measured so that they have credibility. There is a tension between these competing goals.

It is important to start at the beginning—with the concept of *R&D expenditures*. According to the *Frascati Manual*—the basic international source of methodology for collecting and using R&D statistics—an organization may have expenditures either within the unit (intramural) or outside it (extramural) (Organisation for Economic Co-operation and Development, 2002a). Each of the major R&D entities—government, academia, and industry—has a mixture of intramural and extramural expenditures. A proper accounting of R&D expenditures requires an understanding of the flow of resources among units, organizations, and sectors.

In practice, these rather straightforward concepts are quite difficult to measure. For example, government generally makes extramural expenditures. Some that are made through direct procurement, grants, cooperative agreements, or other financial instruments are quite easy to measure and to be reported by the recipient. However, there is an entire class of expenditures in the form of free services, implicit rent, forgiven loans, bonuses, and tax credits for which the value is less directly measurable and much more difficult for the beneficiary to report.

Similarly, industry makes direct investment in R&D that is relatively easy to measure and indirect investment that is harder to quantify. Industrial R&D spending data are collected at the corporate level and classified by the principal business of the firm. Thus, the diversity of R&D performed by many multinational companies is obscured, and large amounts of R&D expenditures can be reclassified from year to year as a result of reclassifying the principal business of one or more large companies (e.g., reclassifying IBM from computer manufacturing to computer services).

Another major issue in the defining of R&D has to do with the general R&D typology—basic research, applied research, and development applied to all sectors—and the imprecision with which these components are reported. For example, there are incompatibilities in the very measure of basic research. The guidance that NSF provides on the questionnaires concerning the cutoff point between basic research and applied research is deliberately vague. Given the diversity and complexity of activities that might be subjectively (and objectively) defined as basic research, the agency has relied on the perspective of the survey respondents in reporting these data. It is admittedly a challenge for most respondents. Even when respondents do have a clear idea of the activities that they would classify as basic research, generally their accounting and budget records are not maintained with this classification coded.

In the survey of business enterprise R&D, basic research is defined to include the cost of research projects that represent original investigation for the advancement of scientific knowledge and that do not have specific immediate commercial objectives, although they may be in the fields of present or potential interest to the reporting company. Applied research is to include the cost of research projects that represent investigation in discovery of new scientific knowledge and that have specific commercial objectives with respect to either products or processes.

For academic R&D, the guidance is to rely on official budgetary and regulatory definitions for R&D: R&D for purposes of this survey is the same as "organized research" as defined in Section B.1.b. of OMB Circular A-21 (revised). It includes all R&D activities of an institution that are separately budgeted and accounted for. R&D includes both "sponsored research" activities (sponsored by federal and nonfederal agencies and organizations) and "university research" (separately budgeted under an internal application of institutional funds). Research is systematic study directed toward fuller knowledge or understanding of the subject studied. Research is classified as either basic or applied, according to the objectives of the investigator. Basic research is directed toward an increase of knowledge; it is research for which the primary aim of the investigator is a fuller knowledge or understanding of the subject under study rather than a specific application thereof. Thus, in reality, basic research is to be defined at the individual grant level by each principal researcher. When this is not possible, each department head or other relevant research coordinator should review the grants. Here is another method used by one institution to estimate the amounts of basic and applied research: all federally funded grants and R&D funded from other universities, foundations, and nonprofit organizations are considered to be basic research. R&D funds received through federal cooperative agreements and federal contracts and most state-funded R&D are, by definition, applied research.

Finally, for the survey of the government sector, NSF and the Office of Management and Budget use the same definitions. Basic research is defined as systematic study directed toward fuller knowledge or understanding of the fundamental aspects of phenomena and of observable facts without specific applications toward processes or products in mind. Applied research is defined as systematic study to gain knowledge or understanding necessary to determine the means by which a recognized and specific need may be met. Development is defined as systematic application of knowledge or understanding, directed toward the production of useful materials, devices, and systems or methods, including design, development, and improvement of prototypes and new processes to meet specific requirements (U.S. Office of Management and Budget, 2003a).

The fields-of-science taxonomies are another example of definitions that have been maintained over time despite clear evidence that the phenomena to be measured have changed. The definition of basic and applied research sponsored by federal agencies and conducted at universities is classified by fields that reflect the traditional—and in the view of some critics—antiquated organization of university departments. The field classification of public-sector research and the industrial classification of private-sector R&D are not comparable, and classifications are not standard across NSF's R&D and personnel surveys. Additional discussion of the fields-of-science issues is found in Chapter 6.

2

Uses and Users

The emergence of the portfolio of R&D expenditure surveys was based on a perceived need for the information. The R&D policies of various administrations and congressional legislation often informed that perception. The result is that the portfolios of recurring and special surveys of R&D expenditures at the start of the 21st century are somewhat eclectic. The surveys range from the flagship federal funds, industrial R&D, and academic spending surveys, which cover the major sectors of R&D activity, to the more narrowly focused survey of federal support for universities, colleges, and nonprofit institutions and the survey of scientific and engineering research facilities at colleges, to the infrequent surveys of innovation, nonprofit institutions, and state R&D spending. Each began with the intent of filling a void in the understanding of the nature and extent of R&D in the United States. Along the way, each has attracted a cadre of users and panoply of uses that extend the utility of the data beyond the original purpose. The users, in the final analysis, determine the relevance of the NSF data offerings.

It is important for NSF to fully understand and respond to the requirements of data users and, to maintain its relevance, to be more anticipatory and proactive in meeting those requirements. As the Committee on National Statistics of the National Research Council (NRC) has advised, a statistical agency's mission should include responsibility for assessing the needs for information of its constituents in order to ensure that the data and information they provide continue to be relevant over time (National Research Council, 2001b).

AN EXPLOSION OF USES

The catalogue of uses and directory of users of the research and development expenditure data of the NSF Science Resources Statistics (SRS) Division has grown by leaps and bounds over its history. Today, SRS data are widely relied on to perform their historical role in measuring federal R&D funding and R&D performance, but increasingly they are asked to serve purposes never envisioned when the data series were initiated. This growth in the community of users and in the variety of uses has outstripped the capacity of the SRS program to provide all the data needed by all users in a fast-changing world. The limitations of the data have prevailed despite the fact that, over the years, the measures have expanded to sharpen the focus on R&D funding in colleges and universities and in the service sectors of the economy.

In order to be more responsive, NSF has attempted to expand the data in depth and detail, but the agency has been only partially successful. For example, the industrial R&D sample was increased to better capture R&D performance in small and nonmanufacturing firms. The addition of state breakdowns is another example. These refinement and catch-up efforts are generally considered to have fallen short of keeping pace with the expansion of requirements for R&D data in the recent past.

At their most basic level, and to a large extent, R&D expenditure data are important in and of themselves. The time series and periodic special studies are critical to the understanding of trends by *source of funding*, *performer of research and development activity*, *type of R&D* (usually following a taxonomy of basic research, applied research, and development), *field of science or engineering*, and *geographic area* (Jaffe, 1996). However, they are incomplete. They measure input, rather than output, of R&D.

Of growing importance, however, are ancillary purposes to which the SRS data have been enlisted. For example, R&D expenditure data today are a key factor in understanding significant processes for which they were not initially designed, such as innovation. Innovation has been defined as the invention, commercialization, and diffusion of new products, processes, and services; these, in turn, are taken to be an important determinant of economic growth, productivity, and welfare (National Research Council, 2001a). The recent NRC report *Using Human Resource Data to Track Innovation* (2001a) points out that R&D expenditure data are often taken as the best surrogate indicator of innovation, in part because of the high degree of industry and firm detail and wide industry coverage and in part because they are the most consistently collected data with annual time series that extend back for decades. Thus, they are taken to represent the best time series related to innovation. Other series, such as counts of patents, counts of contractual collaborations, counts of inventions, and counts of literature citations, are helpful but have problems in duration, coverage, and consistency that reduce their value for understanding innovation trends.

MEETING THE MULTIPLE NEEDS OF MULTIPLE DATA USERS

The long and growing list of uses and users indicates the difficulty faced by SRS in meeting its obligation as a provider of statistical information. This obligation is best depicted in the NRC publication *Principles and Practices for a Federal Statistical Agency* (2001b). In brief, the "best practice" guidance suggests that NSF plan its program and identify emerging issues of import by working "closely with policy analysts in its department, other appropriate agencies in the executive branch, relevant committees and staff of the Congress, and appropriate nongovernmental groups" (p. 3).

For SRS, this is no simple task. The list of constituents that they identify in their publications includes policy makers in the Executive Office of the President, particularly in the Office of Management and Budget, and in the Office of Technology and Science Policy. It includes a number of congressional committees and staff, highlighted by the Senate Committee on Commerce, Science, and Transportation; the House Committee on Science; the Joint Economic Committee; and the Congressional Research Service. The National Science Board and officials in other federal agencies utilize the data in policy formulation, while the statistical arms of the departments of Commerce and Labor use the data as input to programs of economic measurement. The policy formulation users include policy makers at the state and local levels who play a role in education and technology-based economic development. They include those who seek to inform policy, such as the National Academy of Sciences and its subordinate organizations, including the Board on Science, Technology, and Economic Policy; the Board on Higher Education and the Scientific Workforce; and the Committee on National Statistics. Various professional associations and think tanks, most prominently the American Academy for the Advancement of Science, have added to the compendium of requirements.

A growing group of hands-on SRS data users include academic administrators in the nation's colleges and universities and planners and policy makers in industry who have much at stake in federal policies and funding programs. Academic researchers seeking to understand scientific processes and explore relevant science and technology policy issues are critically important data users. The media are also a constituency, as they serve as one means for disseminating data and issue-oriented analysis to the SRS audience. Students and faculty are SRS customers; the more informed their career and mentoring decisions, the more effective the science and engineering enterprise. And finally, international organizations such as the United

Nations and the Organisation for Economic Co-operation and Development heavily utilize the SRS data for international comparative studies.

USER REQUIREMENTS AND PRIORITIES

It is not possible, in this study, to fully explore the dimensions of the requirements imposed by this impressive variety of uses and users, nor is it possible to rank the uses and users. Nonetheless, it is important to capture some of the most crucial requirements for the consideration of the panel and to suggest some that should be addressed as important priorities.

This survey of uses and users does not need to begin at the beginning. Previous studies have concluded that SRS has served several of its multiple user-constituents reasonably well. Representatives of the user community who participated in focus groups and structured interviews for the study *Measuring the Science and Engineering Enterprise* generally believed SRS is already doing an excellent job (National Research Council, 2000). One individual remarked that, overall, SRS data are the "gold standard" for data related to science and technology, noting that coverage of issues is very good. Another interviewee said that SRS data are very helpful for U.S. science and technology issues: "Their domestic data are essential; they are the official statistics. Their international publications are also useful. The data in such publications as *Science and Engineering Indicators*, *National Patterns of R&D Resources,* and *Federal Funds* are quite detailed and very useful." Others were not so complimentary, pointing out gaps in the data, problems of timeliness that rendered the data less useful, and concerns about the data that stem from the discrepancy between funding reported by federal providers and R&D performers.

THE USERS

National Science Foundation

A strategic overview should start at home and focus on two questions: What are the requirements of the National Science Foundation and the National Science Board? How well does SRS meet those needs?

The raison d'etre for the SRS data is to serve the mission of the National Science Foundation. The NSF mission is "to promote the progress of science; to advance the national health, prosperity, and welfare; and to secure the national defense." More specifically, the statistical purpose of NSF, as clarified over the years, is to provide a central clearinghouse for the collection, interpretation, and analysis of data on scientific and engineering resources and to provide a source of information for policy formation by other agencies of the federal government. This charter unambiguously anchors the SRS data in the mission realm of NSF (albeit a wide-ranging realm), while directing the attention of the statisticians to meeting the needs of other *federal* agencies. As the programmatic interests of NSF broadened with the passage of time, so did the interpretation of the scope and function of the SRS R&D expenditure surveys. Today, the focus is broadly understood and accepted to be on policy makers, managers, educators, and researchers in the science and technology arena. Nonetheless, it is useful to go back to the origins.

The expanding scope and intent of NSF interests have largely been responsible for defining today's portfolio of SRS R&D expenditure surveys. The initial data collection was to answer the important but still-difficult question: What is the federal government investing in R&D and how is it performing? The first data collections in the early 1950s and the Survey of Federal Funds for Research and Development, which collects data on *obligations* made by federal agencies, were designed to respond to that narrow inquiry. The question was soon enlarged to permit evaluation of the federal role in view of the total investment in R&D, and the industrial R&D survey was born to answer that question. The data collection net was cast wider over the next two decades, still focusing largely on measuring the federal portion of a total national investment in R&D. Surveys of R&D performance in broadly defined nonprofit institutions, and later, more specifically at major universities, followed. Subsequent additions to the portfolio have deepened and otherwise enhanced the R&D data, keeping in sight the seminal need to better understand the role of the federal government in supporting research and development. There is a tension in the SRS division as it seeks to fulfill these NSF-related objectives of the SRS R&D expenditure data while being responsive to the larger community of users when considering issues of data scope, coverage, and quality.

National Science Board

Within the NSF family, the National Science Board (NSB) is another heavy user of SRS R&D expenditure data. The range of NSB's interests, for example, can be gleaned from a content examination of its biennial report, *Science and Engineering Indicators,* for which the SRS staff has primary production responsibility. The intent of this volume is to "provide a broad base of quantitative information about U.S. science, engineering, and technology for use by public and private policymakers" and "because of the spread of scientific and technological capabilities around the world . . . a significant amount of material about these international capabilities" (National Science Foundation, 2004:iii). The content reflects the wide-ranging interest of the board.

The 2002 volume of NSB's *Science and Engineering Indicators* (hereafter referred to as *Indicators*) includes material on science, mathematics, and engineering education from the elementary level through graduate school and beyond; the scientific and engineering workforce; U.S. and international performers, activities, and outcomes; U.S. competitiveness in high technology; public attitudes and understanding of science and engineering; and the significance of information technologies for science and for the daily lives of U.S. citizens in schools, the workplace, and the community.

Many of the SRS R&D expenditure data series are considered important by the NSB and find their way into *Indicators*. The 2002 *Indicators* report included data from each of the R&D surveys, as well as several outside sources.

In testimony before the panel, the vice chair of the National Science Board, Anita K. Jones, strongly emphasized the role of the R&D data and the *Indicators* publication in national science and technology policy making. She focused on the needs of two groups of policy makers: political appointees in the R&D arena at the federal and state levels, as well as scientists and engineers on advisory boards. These boards are of several types. They include the presidential boards, such as the National Science Board and the President's Council of Advisors on Science and Technology, the NSF directorate advisory committees, agency advisory committees, and National Academies committees. She pointed out that *Indicators* was a leading resource for an R&D policy maker, in that the data are sound, the definitions and categories are stable, and the longitudinal data provide a useful perspective. In particular, she took issue with the recommendation in a previous National Research Council report, *Measuring the Science and Engineering Enterprise,* which recommended making *Indicators* "smaller . . . and less duplicative of other SRS publications." Her testimony urged that a smaller *Indicators* would not be in the best interest of the audience because *Indicators* is the "one-stop shop" for policy makers who do not study the multitude of R&D statistics publications (Jones, 2003).

Office of Management and Budget

SRS data are widely used by officials in the U.S. Office of Management and Budget (OMB). In turn, OMB publishes data on R&D that are not directly comparable to the NSF data. OMB data are based on federal budget authority by functional category, while NSF data report federal obligations. OMB data also include budget authority for R&D plant, a category that is not included in the NSF data.[1] The primary coin of the

[1]Interview with David Radzanowski and David Trinkle, Office of Management and Budget, January 14, 2003.

BOX 2-1
Categories of R&D Spending

Budget Authority. The authority provided by federal law to incur financial obligations that will result in outlays.

Obligations. The amounts for orders placed, contracts awarded, services received, and similar transactions during a given period, regardless of when funds were appropriated or payment required.

Outlays. The amounts for checks issued and cash payments made during a given period, regardless of when funds were appropriated or obligated.

realm for OMB is "budget authority," in that OMB manages and reports the federal budget-by-budget authority classifications. It is important to understand the distinction between the various measures of R&D activity: budget authority, obligations, outlays, and expenditures (see Box 2-1).

In managing the budget of the federal government, OMB has historically divided R&D budget authority into three categories: basic research, applied research, and development. These classifications are coordinated with the National Science Foundation and are developed with a view toward compatibility with international standards and definitions (see Box 2-2).

BOX 2-2
Office of Management and Budget Definitions

Research and Development Activities. Creative work undertaken on a systematic basis in order to increase the stock of knowledge, including the knowledge of man, culture, and society, and the use of this stock of knowledge to devise new applications.

Basic Research. Systematic study directed toward fuller knowledge or understanding of the fundamental aspects of phenomena and of observable facts without specific applications toward processes or products in mind.

Applied Research. Systematic study to gain knowledge or understanding necessary to determine the means by which a recognized and specific need may be met.

Development. Systematic application of knowledge or understanding, directed toward the production of useful materials, devices, and systems or methods, including design, development, and improvement of prototypes and new processes to meet specific requirements.

More recently, additional classifications of the allocation of research funds have been added: congressional direction; inherently unique; merit reviewed with limited competitive selection; merit reviewed with competitive selection and internal evaluation; and merit reviewed with competitive selection and external (peer) evaluation (U.S. Office of Management and Budget, 2003c). OMB also collects and publishes budget authority data on cross-cutting issues identified by the National Science and Technology Council; the U.S. global change research program; networking and information technology; and the national nanotechnology initiative.

The R&D budget authority data published by OMB in the president's budget request are closely monitored by the R&D community. The regular analysis of these data in the president's budget request by the American Association for the Advancement of Science is documented below.

There are several issues that recur in assessing the validity and reliability of the OMB budget authority data. The primary issue has to do with the difference between budget authority and expenditures. Budget authority may not always be translated into spending. For many reasons, programs develop on schedules that deviate from those envisioned in the president's budget, so authority may not translate into spending. Similarly, for many capital programs, budget authority may overlap fiscal years, so the relationship between authority and spending in any single year may deviate considerably. There is also some evidence that, despite common definitions, various federal agencies have different classifications of similar activities, and these differing classifications have most influence on the division between basic and applied research. Finally, it is difficult to compare levels of effort from year to year. Programs move through the sequence from basic to applied research, to development, and then to implementation over the course of several years. As major programs like the space station move through this cycle, the data may change dramatically.

Office of Science and Technology Policy and President's Council of Advisors on Science and Technology

Use of the NSF data by the Office of Science and Technology Policy (OSTP) varies in keeping with evolving emphasis of this agency. In the recent past, direct utilization of NSF data has been limited but, through the facilities of the National Science and Technology Council, for which it has day-to-day operating responsibility, and of the President's Council of Advisors on Science and Technology (PCAST), OSTP turns out to be an extensive user of NSF data. A recent example of the use of the SRS expenditure data is to be found in the PCAST report *Assessing the U.S. R&D Investment* (President's Council of Advisors on Science and Technology, 2002).

The report relies on both budget authority and expenditure data to spotlight trends in R&D investment by the public and private sectors and the shift of emphasis from government to industry investment, which would, in turn, have the consequence of shifting focus from research to development investment.

Congress

Two authorizing congressional committees make extensive use of SRS R&D data in their work. The Senate Committee on Commerce, Science, and Transportation, and the House of Representatives Committee on Science have jurisdiction over nondefense federal scientific research and development programs, and specifically over the programs of the National Science Foundation. SRS expenditure data are often used in committee reports and in testimony before these committees. For example, the House Science Committee's 1998 report, *Unlocking Our Future: Toward a New National Science Policy*, directly referred to a number of data series from the R&D surveys and extensively quoted expert testimony that, in turn, drew heavily on the R&D data (U.S. Congress, House, 1998). In testimony before our panel, David Goldston, chief counsel of the Senate science committee, supported a robust program of R&D data, explaining that the current data are heavily relied on to support congressional decision making.

The Joint Economic Committee has utilized NSF R&D expenditure data in several studies over the past several years that have examined the role of R&D and investment in economic growth and prosperity. Two major studies in 1999, *The Growing Importance of Industrial R&D to the U.S. Economy* and a report on a national high-technology summit, *American Leadership in the Innovation Economy*, built their analysis on extensive use of SRS data (U.S. Congress, Joint Economic Committee, 1999a, 1999b). More recently, SRS data on R&D in information technology were utilized as the basis for a report, *Information Technology in the New Economy*. The data types used in these reports include federal and nonfederal R&D as a proportion of gross domestic product (GDP), R&D by major industry group, and international comparisons of R&D expenditures.

The Congressional Research Service (CRS) often relies on the R&D survey data to answer congressional requests. For example, at the request of the Senate Commerce, Science, and Transportation Committee, the CRS conducted a workshop the addressed the collection and reporting of federal R&D funding data to NSF for its various R&D funding surveys. The report, *Challenges in Collecting and Reporting Federal Research and Development Data* (Congressional Research Service, 2000), remains one of the principle sources of independent assessment of the scope, content, and quality of the SRS data. The report summarized the different aspects of the various disparities that occur in the reporting of federal R&D data by NSF and examines central issues regarding the collection and reporting of the data. The report points out several systemic problems with the collection of R&D data from federal agencies, concluding that reporting R&D data is a burden with little benefit to the agency (Congressional Research Service, 2000). The report also suggests that the fields-of-science categories have become less representative over time because of the changing nature of research. The age-old dichotomy between maintaining currency and maintaining a historical time series comes into play for SRS when confronting this and other issues involved in modernizing the data series.

The CRS use of the federal funds survey was discussed in a November 1998 NSF agency workshop on federal R&D. While expressing general satisfaction with the data, the CRS representative requested a breakdown of federal R&D funding by congressional district. He also urged presentation of data by budget authority rather than outlays and obligations, and in a manner compatible with the organization of the congressional appropriation process, that is, with data disaggregated by the nine congressional appropriation categories. Without such breakouts, R&D funding is forced to compete with other budget priorities rather than to be considered as a separate class of spending across the appropriation categories. Finally, he suggested that the data would be more useful if they were more timely (Quantum Research Corporation, 1999:4-5).

The U.S. General Accounting Office (GAO) uses SRS R&D data to respond to requests for information from members and committees of Congress. In May 2001, GAO studied the reported gap between NSF data reported on obligations of federal agencies for R&D support and the amount that performers (including industries, universities, and other nonprofit organizations) reported spending (U.S. General Accounting Office, 2001). Congress was concerned about this gap, which stemmed, in the view of GAO, from comparing two dissimilar types of data, not from poor-quality data, nor from a systemic problem in receiving and spending federal funds.

Universities

University administrators are also perennial users of SRS data. Administrators use SRS data on university R&D and academic facilities, primarily as a basis of comparison of their own programs with those of their peers and competitors. These uses are particularly important in public colleges and universities, where state legislative oversight is often aided by these comparative data. In SRS surveys for which university administrators are respondents, SRS obtains very high response rates—around 95 percent—because these administrators, in turn, use the data.

Other Organizations

SRS data show up prominently in reports prepared by various other organizations. The American Association for the Advancement of Science primarily uses data from the OMB and from federal agencies in its annual report on federal research and development spending in the president's budget. NSF data are used to provide a historical context for these more current data. In a report on the "new economy" prepared by the Progressive Policy Institute, 5 of 39 indicators drew on SRS data from either *Science and Engineering Indicators* or *National Patterns of R&D Resources*, compared with three drawn from the *Economic Report of the President* and seven drawn from Bureau of Labor Statistics data (Atkinson and Court, 1998). Likewise, the Committee for Economic Development recently released a report, *America's Basic Research*, in which almost 90 percent of the data in the report's tables and figures were SRS data drawn from either SRS publications or NSB's *Science and Engineering Indicators* (Committee for Economic Development, 1998).

Industrial R&D Research Analysts

Over the past decade, the industrial R&D research community has focused increasingly on understanding how firms organize to perform R&D, with the ultimate goal of identifying types of operations that achieve higher productivity. This may be considered a micro-level version of the work being conducted by the Bureau of Labor Statistics (BLS). For these purposes, data with considerable detail are needed—information on connections of firms with other firms, with academia, and with government laboratories. The structure of the R&D workforce within an organization is also of interest, with a focus on degree levels and degree fields employed in the R&D venture. The SRS staff has responded by increasing the amount of detail, although mindful to balance the increase in detail with the increase in respondent burden. Beginning with data for 2001, the NSF industry R&D survey will collect detail on extramural R&D (R&D performed outside the company but inside the United States) by type of contract organization: for-profit, university, and nonprofit. The academic R&D expenditure survey initiated collection of pass-through R&D funding, that is, total dollar amounts sent to (starting in 1996) and received from (starting in 2000) higher education and all other institutions. The one-time information technology innovation survey, in progress this year, delves into these issues in considerable depth.

Consistent with this demand for more internal detail on organizations, academic R&D researchers have urged SRS to collect data at lower levels of aggregation—plants, not firms—and by line of business rather than SIC major industry grouping. This requirement for fine-sizing of the data has significant implications for the sample size and the sample frame selected for the survey and, in consequence, on the overall design and cost of the surveys.

Over the years, academic researchers and analysts have made significant contributions to both the understanding of the R&D enterprise and its measurement. In addition to sorting out the role of R&D in productivity measurement, contributions have been notable in conceptualizing and measuring innovation, in understanding the new organizational structures for the conduct of R&D, and in focusing attention on the international aspects of R&D. The academic research community has been instrumental in illuminating avenues of inquiry and in challenging NSF to improve the measures.

Other Potential Users

The panel thinks that there is the potential to expand the demand for SRS data in the private sector. Members of the industrial business community could become SRS data users if weaknesses in data publication are addressed. In particular, business firm representatives to the Industrial Research Institute have indicated that, if the industrial R&D data were reported on a more timely basis and were disaggregated by line of business, they would use the data for benchmarking. Response rates and accuracy might well improve if respondents to the industrial R&D survey think that the data can be of some use to them.

OTHER GOVERNMENT USES

National Income and Product Accounts

A company could build a new plant to produce one of its products, so a tracking of company expenditures would note the new investment, and analysts could relate the new investment to the production of the product. Such investments are captured by the Bureau of Economic Analysis (BEA) in its System of National Accounts. If a company invests in new plants for two distinct products, BEA would prefer that data collection activities track the investments and their effects separately for the two products.

Since the Census Bureau collects data for the NSF industry survey at the operating establishment level, much of the need at BEA for separate data on separate products is satisfied. Both production data for the industry R&D survey and capital expenditures data are available at the five-digit NAICS (four-digit SIC) industry level.

By extension, national income and product account analysts would like R&D investment data at the same level of detail and collected in much the same manner. Such an approach would enable them, for example, to put the replacement of an old machine with a new one, and the replacement of an old way of using the machine with a new method, on the same basis. No such data are collected. Companies do not report R&D investment data to any government agency, nor do they report R&D expense data for separate industries. There are serious doubts about whether all R&D expenditures can be separated at the 5-digit NAICS detail, particularly expenditures for basic research, which may benefit several detailed industry groups.

> In BEA's input-output accounts, neither current expenses nor receipts for R&D are identified at the published level of detail. A portion of R&D is identified at the level of detail at which the estimates are prepared (Bureau of Economic Analysis, 1994:42, fn. 17).

BEA has extensively utilized SRS data in developing a "satellite" account for research and development to supplement the existing national accounts. First introduced in 1994 (Bureau of Economic Analysis, 1994), the satellite account provides estimates of expenditures on R&D that are used in conjunction with the national income and product accounts. The satellite account treats R&D expenditures as a form of investment, recognizing the role R&D plays in adding to knowledge and in developing new and improved processes and products that lead to increases in productivity and growth. The satellite account provides estimates of the stock of knowledge capital.

The estimates of R&D expenditures in the satellite account are based on four of the SRS surveys: Federal Funds for Research and Development, Federal Science and Engineering Support to Universities, Colleges, and Nonprofit Institutions, Research and Development Expenditures at Universities and Colleges, and Industrial Research and Development. The accounts also used input from the SRS surveys of state and local and nonprofit institution R&D expenditures. Several significant adjustments to the SRS data were required to make them malleable to the national accounts, including removal of expenditures for R&D structures and equipment, converting fiscal to calendar years and federal obligations to expenditures, and, importantly, substituting judgmental estimates for R&D data that have been suppressed by NSF to avoid disclosure of confidential data.

A more recent study (Fraumeni and Okubo, 2002) has explored options for going a step further to actually include R&D in the official national accounts. The proposal to capitalize R&D expenditures would raise investment on the product side of the accounts, thus raising GDP, while changing its composition (investment rises and consumption falls while the level and rate of savings increase). In proposing these changes, the authors recognize several key limitations in the underlying SRS R&D expenditure data for this expanded purpose. These recognized limitations lead to the expectation that, if this important revision to treatment of R&D in the national accounts were introduced, it would greatly increase the visibility of the SRS data series, putting additional pressure on the quality and timeliness of the data.

Productivity Estimates

Good productivity estimates need good estimates of R&D. Much of what economists know empirically about the relation between R&D and productivity is based on studies that regress data at the company level on the R&D numbers along with ratios of other inputs (labor, tangible capital, etc.) (Griliches, 1980). The model is forced to assume that each firm operates in a single industry.

Studies that have used lines of business as individual observations have not as a rule recalculated regression equations after aggregating up to the company level. Those that have used company-level data and would have used line-of-business data if they had had them typically do not know enough to model the regressions to ameliorate the negative econometric effects. For studies of the effects of R&D on productivity, the general effect may be to show a smaller effect or a less significant effect.

Multifactor Productivity Estimates

The Bureau of Labor Statistics utilizes SRS R&D expenditure data in development of estimates of productivity in the nonfarm economy. The data are used as input to the BLS annual multifactor productivity report (U.S. Bureau of Labor Statistics, 2002). In the computation of multifactor productivity, the stock of research and development in private nonfarm business is derived by cumulating constant-dollar measures of research and development expenditures and allowing for depreciation. The current dollar expenditures levels for privately financed R&D are obtained from the NSF industry R&D series, while price deflators and estimates of depreciation are developed by BLS.

The BLS data on multifactor productivity deal only with the direct return to research and development, that is, productivity gains by industries identified by the amount of R&D they do for themselves. The indirect effects of research and development obtained by purchasers further along the chain of production are likely to be significant. In order to flesh out these series, BLS has conducted research in utilizing NSF data to identify the "indirect effects" of R&D by constructing indexes of the R&D intensity of product producers and using BLS capital data and input-output tables to regress the productivity of downstream users of these products on R&D intensity. As reported in a 1989 study, *The Impact of Research and Devel-*

opment on Productivity Growth, in order to better identify and break down these indirect effects, BLS would prefer to use data by detailed industry, much like the line-of-business series produced by the Federal Trade Commission, determine the precise effect of federal R&D expenditures on private nonfarm productivity (U.S. Bureau of Labor Statistics, 1989).

Occupational Employment Projections

The Bureau of Labor Statistics also makes extensive use of the NSF industry data in its program of occupational employment projections. BLS staff report that more extensive use of the data would be made if the NSF data (1) provided a measure for total technology-oriented resources; (2) were further disaggregated by industry category; and (3) were a measure of output rather than intensity of effort. Despite these shortcomings, it is expected that BLS use of the NSF data will grow in the future because a competing series derived from the Occupational Employment Survey has been discontinued.

Program Management

Integration of R&D activities into the national accounts and for productivity estimates are not the only interest in R&D at the national level. Both tax credit and subsidy policies are important, and the details in each of those areas can depend on differences across industries. Currently not enough reliable data on industries are available to enable policy makers to tailor tax credits by industry. Subsidies are typically keyed to broad industry categories: the Defense Advanced Research Projects Agency targets defense applications, grants flow to specific broad areas via appropriations to the National Institutes of Health and the departments of Energy, Transportation, and Homeland Security. Even general-purpose grant programs like NSF and the National Institute of Standards and Technology's Advanced Technology Program are organized along fairly specific technology areas.

3

Measuring R&D in Business and Industry

The annual survey of R&D in industry is one of the most important products of the NSF's Science Resources Statistics (SRS) division. It measures a sector that accounts for the bulk of R&D activity in the economy, accounting for 66 percent of funding for R&D by source and more than 70 percent of funds spent on R&D.

Because of the predominance of the industrial sector in R&D, the industrial R&D expenditure data are closely watched as an indicator of the health of the R&D enterprise in the United States. For example, a recent downturn in real spending on industrial R&D has given rise to some concern over the health of that enterprise.[1] The decline in industrial R&D spending between 2001 and 2002, coinciding with the overall economic slowdown, was identified by NSF as the largest single-year absolute and percentage reduction in current-dollar industry R&D spending since NSF initiated the survey in 1953 (National Science Foundation, 2004). This downturn was identified as the leading cause of the downtick in the ratio of R&D spending to gross domestic product from 2000 to 2002.

This most important R&D survey is also the most problematic. No survey in the NSF portfolio of R&D expenditure surveys has been as impacted by changes in the R&D environment (discussed in Chapter 1) as the Survey of Industrial Research and Development.

This survey is also the most sensitive to changes in the procedures for statistical measurement. Little is known about respondents, what they

[1]Industrial R&D failed to keep pace with inflation and experienced its first decline in real terms after 1994 (National Science Foundation, 2002).

know, and how they treat this survey. We do not know who is responding in each firm, or what their position or capabilities are. Nor do we know how the data each firm reports are gathered within the firm, or their relationship to similar data reported to the Securities and Exchange Commission (SEC).[2]

Since its inception in 1953, the survey has experienced several focused efforts to modernize and strengthen its conceptual and technical foundation. The most ambitious modernization effort took place a little over a decade ago, in 1992, when the sample design was substantially modified to better represent the growing service sector, extend coverage to smaller firms, and collect additional information for understanding R&D outsourcing and that performed by foreign subsidiaries.

The 1992 redesign effort updated the operations of the survey to reflect many aspects of the changed environment and correct some of the most severe deficiencies, which had rendered the survey results quite misleading over time. At the conclusion of that redesign, however, there was still much to do to meet a growing need for data to measure the emerging realities in the conduct and measurement of R&D in the United States. Not much progress has been made since that redesign; in fact, the survey changes that have been introduced since 1992 have been almost cosmetic in their application. Minimal revisions have been implemented to maintain currency, such as the introduction of the North American Industry Classification System (NAICS) in 1997 and the more recent changes in questionnaire wording and collection procedures (see Box 3-1).

Thus, this decade-old redesign of the survey has left several issues unresolved. A major unresolved issue is the failure to implement collection of R&D data at the product or line-of-business level of detail; another is the failure to learn more about respondents and to sharpen concepts and definitions to more adequately reflect business organization for R&D; and a third is the inability to speed the production and release of the estimates to data users.

With the passage of time, the need to consider further revisions has accelerated. The R&D environment has changed even further since the early 1990s. As discussed in Chapter 1, organizational arrangements for the conduct of R&D have continued to change apace. New forms of R&D in

[2]Hall and Long (1999) compared data from the 1991 and 1992 Survey of Industrial Research and Development to the SEC 10K reports and found a number of discrepancies of three basic types: differences between fiscal and calendar year reporting, differences in coverage (whether foreign-performed R&D was included), and differences in definition, either intended or unintended. They pursued the source of the definitional differences in interviews with a small number of firms. Given the confidential nature of the RD-1 data, detailed tables of the discrepancies by industry and reason could not be produced, so that a precise picture of the relationship between SEC and NSF data was not obtained.

BOX 3-1
Thumbnail History of the Industrial R&D Survey

1953 First Survey of Industrial Research and Development is conducted by the Bureau of Labor Statistics.
1957 The Bureau of the Census assumes responsibility for the survey.
1967 The sample design called for the selection of a new sample (about 14,000 companies) every 5 years with estimates updated from a subset or panel of (about 1,700) companies in the intervening years. The current-year panel was a subset of the sample drawn for 1967.
1976 Sample wedging was introduced to update the sample between the irregular interval sample selections.
1992 This was the first year of "annual" sampling. Beginning with 1992, a new sample of 23,000-25,000 firms was selected annually. Previously, the sample frame was limited to companies above certain size criteria based on number of employees that varied by industry; for 1992 and beyond these size criteria were dropped. Sampling error maximums were established, 2 percent for industries in which there was indication of a high level of R&D activity, 5 percent for all others. And 25 new nonmanufacturing industry groups were added to the sampling frame.
1994 Criteria for predetermination of "certainty" selections for the sample were changed. The predetermination was limited to companies with reported or estimated R&D expenditures of $1 million or with 1,000 or more employees. Also, the sampling frame for each industry sampling recode stratum was partitioned into large and small companies based on payroll, thereby expanding the use of the more efficient simple random sampling technique for the majority of companies. The probability proportionate to size sampling technique was used for the balance of the frame.
1995 Sampling strata were redefined to correspond to the industry or groups of industries for which statistics are published.

the service sector (particularly in the biomedical fields), the expanding role of small businesses, the geographic clustering of R&D, cooperation, collaboration, alliances, "open innovation," partnerships, globalization, government programs to promote advanced technology demonstrations and nanotechnology—all have impacted the relevance and quality of the data. The list of structural and organizational changes in the business world that impact on the ability to understand the extent and impact of R&D grows daily. Despite NSF staff efforts, the survey is not keeping up.

The panel concludes that it is time to implement another major redesign of this survey (Conclusion 3.1). The redesign would take a four-pronged approach:

- The redesign would begin with a reassessment of the U.S. survey against the "standard," that is, the international definitions as promulgated

1996	The predetermination of certainty selections for the sample was limited to companies with reported or estimated R&D expenditures of $5 million or more. Previously, the criteria were R&D expenditures of $1 million or more or 1,000 or more employees. Instead of being based on payroll, partitioning of the sample was based on employment. Manufacturing (< 50 employees) and nonmanufacturing (< 15 employees) small company totals were collected based on simple random sampling and reported with the "other industries" residual. A "zero industry" designation was established for those industries that reported no R&D expenditures in 1992-1994 but may report R&D in the future.
1999	The North American Industry Classification System (NAICS) was introduced into the survey, and industry codes assigned during the sampling process were retained for publication of the resulting statistics. No other major revisions to the sample design were made since the previous survey cycle. This was the first year that the small company totals were reported as a separate "small companies" category.
2002	For the first time in the survey's 50-year history, the full survey was mandatory for this single survey cycle. The sample size was expanded to improve state estimates. Sampling, estimation, and reporting of small company strata were eliminated since analytical interpretation of these cells was problematic. Industry Federally Funded Research and Development Centers were not included in the survey estimates.
2003	An R&D check-off item was placed on the Census Bureau's Company Organization Survey, an annual survey sent to all multi-unit companies that is used to update the Standard Statistical Establishment List (SSEL)—the frame for the industry R&D survey in the 2003 survey statistical period. The inclusion of this question was tested as a way to create an R&D registrar.
2004	A redesigned questionnaire was introduced based on cognitive research. The major impact was to put open-ended items into question format.

through the *Frascati Manual* (Organisation for Economic Co-operation and Development, 2002a), thereby adding some data items. It would benchmark U.S. survey methodology against best practices in other countries, many of which appear to be producing data of better quality and with more relevance than NSF.

- In order to sharpen the focus of the survey and fix problems further identified in this report, the redesign would update the *questionnaire* to facilitate an understanding of new and emerging R&D issues. In particular, it would test and implement the collection of data on R&D funds from abroad, from affiliate firms, and from independent firms and other institutions for the performance of R&D in the United States. It would sharpen the question on the outsourcing of R&D to distinguish between payments to affiliated firms, to independent firms, and to other institutions abroad. It would make more extensive use of web-based collection technology after appropriate cognitive and methodological research.

- The redesign would enhance the program of *data analysis and publication* that would facilitate additional respondent cooperation, enhance the understanding of the industrial R&D enterprise in the United States, and provide feedback on the quality of the data to permit updating the survey methodology on an ongoing basis. The redesign would be supported by an extensive program of research, testing, and evaluation so as to resolve issues regarding the appropriate level at which to measure R&D, particularly to answer, once and for all, the question about the collectability of product or line-of-business detail.
- The redesign would revise the *sample* to enhance coverage of growing sectors and the *collection procedures* to better nurture, involve, and educate respondents and to improve relevance and timeliness.

To assist NSF in identifying these venues for improvement and in prioritizing these tasks, this chapter addresses each of these areas.

INTERNATIONAL STANDARDS

The concepts and definitions of R&D used in the United States today are generally in keeping with international standards. This is not surprising. After all, the science of measuring R&D was pioneered in the United States, and it has proceeded ahead by virtue of an international effort marked by remarkable comity and underscored by collaborative development. This collaboration among the national experts in member countries of the Organisation for Economic Co-operation and Development (OECD) who collect and issue national R&D data has been ongoing for about five decades, under OECD auspices. The collaboration has been codified in the *Frascati Manual*, which was initially issued about 40 years ago, with several updates over the years, the most recent in 2002.

Nonetheless, there is increasing evidence that the United States is departing somewhat from the reporting standards used internationally. For example, the OECD publication *Main Science and Technology Indicators* (Organisation for Economic Co-operation and Development, 2004) provides an indication that some data are not comparable within OECD guidelines. Although not of serious consequence, these departures indicate that some catch-up work may be needed in order to conform with international standards.[3]

[3]The R&D expenditure data for the United States are underestimates for a number of reasons: (1) R&D performed in the government sector covers only federal government activities. State and local government establishments are excluded; (2) In the higher education sector, R&D in the humanities is excluded, as are capital expenditures; and (3) R&D expenditure in the private nonprofit (PNP) sector covers only current expenditures.

Progress in Other Countries

In spite of a long history of U.S. leadership in measuring R&D, there is evidence that other countries may have caught up with and moved ahead of the United States in the measurement of industrial R&D. To the extent that this may have happened, there are reasons for the trend. Other countries generally have more access to administrative sources of data, or they have traditions in statistical collection that allow (or prescribe) more intrusive collection of data in greater depth than is considered feasible in the United States. The experience of countries in the European Community with the development of the round of Community Innovation Surveys (CIS) is instructive in this regard. The fact that other countries have developed more information on foreign funding of domestic R&D is an indicator of how they have extended knowledge of the R&D enterprise beyond that available to U.S. policy makers.

One country that has made recent strides in the measurement of industrial R&D, which could well be emulated in many respects in the United States, is Canada. It has developed a robust R&D expenditure measurement program.

The Industrial Research and Development Survey in Canada

A survey of research and development performance in commercial enterprises (privately or publicly owned) and of industrial nonprofit organizations has been conducted in Canada since 1955. The survey has changed in frequency, in detail, and in use of administrative data over the years. Now it is an annual survey of enterprises that perform $1 million (Canadian) worth or more of R&D, complemented by the use of administrative data from the tax authority in order to eliminate the reporting burden for small R&D performers. The tax data arise because of the federal Scientific Research and Experimental Development program, which, in some form, is

---

Depreciation is reported in place of gross capital expenditures in the business enterprise sector.

Higher education (and national total) data were revised back to 1998 due to an improved methodology that corrects for double-counting of R&D funds passed between institutions.

Breakdown by type of R&D (basic research, applied research, etc.) was also revised back to 1998 in the business enterprise and higher education sectors due to improved estimation procedures.

Beginning with the 2000 Government Budget Appropriations or Outlays for R&D (GBAORD) data, budgets for capital expenditure, "R&D plant" in national terminology, are included. GBAORD data for earlier years relate to budgets for current costs only.

The United States Technological Balance of Payments (TBP) data cover only "royalties and license fees," which are internationally more comparable. Other transactions, notably "other private services," have been excluded.

available to all firms that perform R&D. The data are available to the statistical office under a provision of the Statistics Act.

Not only are the tax files a source of data for R&D performing enterprises, but they also provide the list of enterprises in Canada that perform R&D. Of course, not all firms apply for the tax benefit, as some regard the administrative demand and the risk of audit as outweighing the benefits. However, these firms tend to be small performers of R&D with few staff committed to the activity. With the survey frame given by administrative records, the statistical office is able to take data directly from the appropriate tax form for about 85 percent of the population of R&D performers. For the rest, those that perform $1 million or more of R&D, a more elaborate questionnaire is mailed to the enterprise, once a contact person is identified as a result of an initial telephone contact. It is felt that this initial contact is essential to the success of the survey.

The information collected includes the expenditure on the performance of R&D, including current and capital costs, and the full-time-equivalent number of personnel engaged in the activity. As well as collecting data on the performance of R&D, questions are asked about the source of funds. There are five sources: governments (federal or provincial); business, including the firm itself; higher education; private nonprofit organizations; and foreign firms and other institutions. About 20 percent of funding for industrial R&D in Canada comes from abroad, as measured in the Canadian R&D survey. In contrast, the United States does not collect this type of information on the industrial R&D survey, missing an opportunity to illuminate this important aspect of the globalization of R&D. Although data exist on foreign sources of R&D funding for other countries, there are no data on foreign funding sources of U.S. R&D performance (National Science Board, 2004).

The country of control of the R&D-performing firm in the Canadian survey is derived from administrative data, which are used to examine the different characteristics of foreign-controlled and domestic-controlled enterprises in relation to their performance of R&D. About 30 percent of the expenditures on industrial R&D in Canada are made by foreign-controlled firms. This spending emanates from within Canada; it is separately identified although counted as domestic R&D spending, in keeping with international practice.

Most R&D performers in Canada have a single head office with one geographical location for both their production and R&D units. However, there are large firms that have production units classified to different NAICS codes, and the production units may be in different provinces. For these firms, special follow-up is required to identify the geographical and industrial allocation of the R&D performed in Canada. Once this is done, the data can be presented by geographical region.

The survey is also a vehicle for questions about new or growing activities, such as R&D in software, biotechnology, new materials, and the environment. It collects data on payments and receipts for R&D and other technological services. And it supports international comparison of Canada's R&D performance (for more details, see Statistics Canada, 2003).

ADEQUACY OF CONCEPTS AND DEFINITIONS

Mainly for purposes of historical and cross-sectional comparisons and continuity, the definitions that underscore NSF's collection of R&D expenditure data from industry have followed a one-size-fits-all philosophy. This approach impedes understanding of the scope and nature of industrial R&D.

The division of industrial R&D activities into the standard categories of basic research, applied research, and development has characterized the industrial R&D survey since its inception. In an effort to maintain a concordance with the definitions collected in federal and academic surveys, as well as with international sources, NSF defines industrial basic research as the pursuit of new scientific knowledge that does not have specific immediate commercial objective, although it may be in fields of present or potential commercial interest; industrial applied research is investigation that may use findings of basic research toward discovering new scientific knowledge that has specific commercial objectives with respect to new products, services, processes, and methods; and industrial development is the systematic use of the knowledge and understanding gained from research or practical experience directed toward the production or significant improvement of useful products, services, processes, or methods, including the design and development of useful products, materials, devices, and systems (U.S. Census Bureau, 2004).

The reporting of data by these basic categories is only as sound as the understanding of the meaning of those categories. In an important 1993 study, Link asked research directors from three research-intensive industries about the accuracy of the categories for describing the scope of R&D that is financed by their companies. Most of the industrial firms reported that the categories described their scope, but there were disagreements among the largest firms. The disagreement had to do with the narrowness of the definition of development. The researcher's conclusion was that the NSF definition understated the amount of development, from the perspective of the firms (Link, 1996).

This problem at the outer edge of the definition of research and development is not surprising since, with some notable exceptions like pharmaceuticals and semiconductors, most corporate investment into R&D exploits science-based inventions that are beyond the research and development phase. The development of new goods and services is often based on costs and specifications defined by marketing opportunities. Branscomb and Auerswald (2002) point out that blurred distinctions between these traditional categories complicate accurate analysis of existing data.

Ambiguous usage of these common categories leaves the door open for variation in interpretation by survey respondents, especially across different firms and industries. Jankowski (2002) points out that the service sector (finance, communications, transportation, and trade) creates value and competes by buying products and assembling them into a system or network, efficiently running or operating the system, and providing services for customers who are often members of the public. It is an open question whether the system design, system operation, and service design and delivery functions can be readily classified into the traditional components of R&D. An example suggested to the panel during its workshop would be a major retailing company that has invested in innovative supply chain management techniques. Such a company would quite possibly not recognize and report this activity as R&D. It was observed that at least one large retailer with widely recognized leadership in distribution management reports no R&D program in its 10K filings to the SEC.

The distinction between applied research and the two other components of R&D is especially problematic in the industrial sector. Applied research could well involve original research believed to have commercial applications, and it could include research that applies knowledge to the solution of practical problems (Branscomb and Auerswald, 2002). Although, the share of R&D represented by applied research has been fairly steady at about 20 percent of the total since the industry data were first published in 1953, the possible blurring of the lines between the basic and applied research and between applied research and development has the potential to distort analysis of the role of R&D in the generation of innovation and growth. These questions are important for focusing and assessing such federal investment programs as the Advanced Technology Program and the Small Business Innovation Research programs, in which the federal government provides project-level support of early-stage commercial technological development.

In assessing the impact of R&D on industries, these distinctions are also important. Branscomb and Auerswald observe that mature industries, such as the automotive sector, tend to invest a smaller percentage of R&D into earlier stages of technological development than do industries at an earlier stage of evolution, such as biotechnology. They make the case for this evolutionary shift in R&D investment strategy by tracing the emerging emphasis on maximizing the "yield from R&D." This has been accomplished by reducing the amount of basic research accomplished in corporate laboratories, outsourcing, collaborations with universities, and developing corporate venture capital organizations that spin off from the core business.

In contrast, the overwhelming focus of U.S. industry on *development* activities more than on pure research means that three-fourths of R&D activity is lumped into these categories that are closer to the customer. These are less likely to be carried out in formal R&D facilities and thus less likely to be classified as R&D. Especially in the service sector, the failure to recognize these customer-driven activities as *development* would tend to bias the estimates of R&D expenditures downward. At the same time, there are classification issues in the development category. Today, companies classify activities that range from advanced technology development to operational system development in the same general category. There is evidence that companies are having difficulty in separating development from engineering as technology advances (McGuckin, 2004). The same issue exists for "technical service," which has certain elements of development embedded. In consequence, the lumping of these activities into one aggregated "development" category tends to obscure important shifts in R&D emphasis over time.

Definitional quirks lead to other anomalies. Software development simulating product performance in lieu of product tests is reported as R&D, but other software development, such as software to predict the nature of consumer demand, is not considered R&D. Heavy investments in technical service developments in quality control and in artificial intelligence and expert systems are similarly uncounted in the R&D expenditure estimates.

The panel learned that companies may be able to provide additional useful breakouts of more detailed information on the components of R&D if the categories are carefully constructed for ease of collection. For example, one large manufacturing corporation maintains detailed records on investment in developmental functions and could report using a classification structure akin to that adopted in the Department of Defense Uniform Budget and Fiscal Accounting Classifications for RDT&E (research, development, test, and evaluation) Budget Activities.[4] It is not certain whether other companies could be similarly accommodating to providing additional detail to help further define the boundaries and details of the R&D classification structure. There is a need for additional classification detail to better describe differences between the industrial sectors and to better focus on the sources of innovative activity.

The potential lack of understanding of the concepts and definitions in the Survey of Industrial Research and Development (often referred to as RD-1, the number on the survey form) is especially troubling. A Census

[4]These activities are (1) basic research, (2) applied research, (3) advanced technology development, (4) advanced component development and prototypes, (5) system development and demonstration, (6) RDT&E management support, (7) and operational system development.

Bureau study found that R&D professionals and support personnel have a good understanding of these concepts, even though they do not necessarily classify their own work that way (U.S. Census Bureau, 1995). However, the persons most often charged with responding to the questionnaire are in the financial or government relations offices of the companies, and their understanding is less likely to be accurate. This finding was confirmed in panel discussions with representatives of reporting companies and covered industries. Although definitions of these terms are included in the instructions transmitted with the forms, prior to the 2003 survey they were not included on the form and thus may have been overlooked. Abbreviated instructions have been included in a revised 2003 questionnaire introduced in March 2004.

NSF should conduct research into recordkeeping practices of reporting establishments by industry and size of company to determine if they can report by more specific categories that further elaborate applied research and development, such as the categories utilized by the Department of Defense (Recommendation 3.1).

ALIGNMENT OF THE SURVEY WITH TAX AND ACCOUNTING PRACTICES

R&D data have never existed in isolation. From their beginning, they were molded and shaped in a larger context. The work that went into developing the collection of information on federal R&D emanated from the concern of Vannevar Bush and others over the lagging role of the U.S. R&D enterprise. Similarly, the concern over appropriate measurement of industrial R&D arose from a concern over the impact of R&D on productivity and economic growth.

By the same token, the conceptual foundation of industrial R&D and the basis for the R&D measures have, for two decades, been closely associated with U.S. tax and accounting policies. Along with international standards and policies promulgated by the U.S. Office of Management and Budget, the tax code and accounting rules have shaped and defined R&D.

A study by Hall (2001) for the European Union finds that there are several features of federal corporate tax law that have an effect on the amount and type of R&D: expensing of R&D; the research and experimentation (R&E) tax credit; foreign source allocation rules for R&D spending; the preferential capital gains tax rate; accelerated depreciation and investment tax credits for capital equipment; and the treatment of acquisitions, especially as it relates to the valuation of intangibles.

The foremost provision of the tax code that defines R&D is the federal research and experimentation tax credit. The credit is designed to stimulate research investment by virtue of a tax credit for incremental research expenditures. The justification for the tax stimulus, as described by the National Science Board (2002), is that returns from research, especially long-term research, often are hard to capture privately, as others might benefit directly or indirectly from them. Therefore, without some policy such as a tax credit, businesses might engage in levels of research below those that would benefit a broader constituency (National Science Board, 2002:4-16). The tax credit is a feature in many developed countries, and it has been a pillar of U.S. government R&D policy since 1981.

In application, the tax credit appears to have a significant impact on corporate R&D accounting and decision making. The credit is provided for 20 percent of *qualified* research above a base amount, so it becomes worthwhile for a company to clearly identify activities that meet the definition of qualified research. Young companies have different, more liberal rules than older companies, so companies may adjust their corporate R&D strategies as they age. Finally, the tax credit has provisions for basic research payments paid to universities and other scientific research organizations, perhaps encouraging collaborative arrangements and outsourcing.

Although the tax credit legislation does not define R&D, the Internal Revenue Service (IRS) has issued defining regulations that define R&E as expenditure incurred in connection with the taxpayer's trade or business that represents *research and development costs in the experimental or laboratory sense*. Research expenses that are qualified for claiming under the regulations are those that include research undertaken for discovering information that is technological in nature, for which the application is intended to be useful in the development of a new or improved business component, and that relate to new or improved function, performance, reliability, or quality. In a series of decisions, tax courts have held that this definition does not include R&D performed by financial service institutions or involving software development more generally.[5]

The tax code goes on to define *basic research* as any original investigation for the advancement of scientific knowledge not having a commercial objective. The definition includes domestic research conducted by foreign firms, but it does not include basic research conducted outside the United States by affiliates of U.S. companies.

Hall (2001) indicates that, in contrast to the tax code, the Federal Accounting Standards Board, which sets the standard for reporting to the Securities and Exchange Commission on the 10K form, defines R&D as follows:

---

[5]*Tax&Accounting Software Corp. v. U.S.*, 301 F.3d 1254 (10th Cir. S002) and *Eustace v. Commissioner*, 2002 U.S. App. LEXIS 25530 (7th Cir. 2002).

> Research—planned search or critical investigation aimed at the discovery of new knowledge with the hope that such knowledge will be useful in developing a new product or service or a new process or technique or in bringing about a significant improvement in an existing product or process.
>
> Development—the translation of research findings or other knowledge into a plan or design for a new product or process or for a significant improvement to an existing product or process.

These tax and accounting definitions are not precisely the same as the international standards and the associated U.S. definitions of the components of R&D. A company is confronted with a multiplicity of definitions and interpretations of activities called R&D that are impacting it at various levels. An activity may be defined as R&D in the laboratory but not by the tax accountant. Neither the laboratory nor the tax accountants may fully agree with the operating definition used by the company agent who has responsibility for completing and submitting the RD-1 form to the Census Bureau.[6]

Research and development spending by businesses plays a significant role in valuing intangibles. Although the tax accounting standards expense R&D, there is frequent discussion over the appropriateness of treating R&D as an investment. The Federal Accounting Standards Board is currently considering proposals to capitalize acquired in-process R&D. Obviously, further moves of regulators in the direction of recognizing intangibles as assets will provide valuable information for the surveys.

In any event, full exploration of the treatment of R&D as an intangible investment requires definitional and measurement breakthroughs. For example, a recent study by Corrado et al. (2004) identified R&D as the largest source of business investment on intangibles. Their taxonomy defines conventional R&D (science and engineering research and development) as the major component of "scientific and creative property." Other components of this subcategory of intangibles are copyright and license costs and other product development, design, and research expenses (particularly in the finance and service sectors). (The latter two subcategories are currently not included in the industrial R&D surveys.) Other intangible assets identified include "computerized information" and "economic competencies" and are clearly not captured by the survey.

Some evidence of the impact of classification of developmental activities and differing treatment of foreign R&D was identified in the 1999 comparison of the SEC 10K reports and the NSF industry R&D estimates

[6]See Altshuler (1988) for an analysis that uses tax data from 10K forms to provide estimates of the share of R&D that is R&E.

by Hall and Long (1999). The RD-1 instructions exclude "routine product testing" and "technical services" from the definition of R&D. Companies have difficulty in distinguishing between nonroutine and routine processes, especially in light of the 10K standard that narrowly defines modifying a product to satisfy a specific customer's needs as a routine customer service to be excluded from R&D, even if the work is performed by R&D employees. Companies thus face two sets of definitions and reporting rules. Differing treatments of foreign R&D also affect the series. Hall and Long found that, in the aggregate, the SEC 10K data do not show a slowdown in the 1980s, while the NSF data exhibit a decline.

In summary, tax and accounting standards play an important role in determining the definitions and reporting mechanisms in firms that report on the industry R&D survey. One may speculate that the substantial scrutiny on corporate bookkeeping in recent years has served to give even more impetus to improvement in reporting than was the case a few years ago when Hall and Long concluded that "substantial effort appears to have gone into getting the R&D numbers 'right' in the professional accounting world, by which we mean following the definitions and reporting requirements carefully and systematically" (Hall and Long, 1999:27).

THE INDUSTRIAL R&D SURVEY

Selecting an Appropriate Scope for the Industrial R&D Survey

The relevance of the industrial R&D survey is defined both by the content of the data collection and by the selection of businesses covered by the survey. This selection is called the "scope" of the survey. The scope has changed over the years to reflect the changing nature of R&D and the changing locale in which R&D is conducted in the United States, but there are still major shortfalls in the scope of the survey (Kusch and Ricciardi, 1995).

The survey has always included manufacturing and most nonagricultural industries. From its earliest years, it has excluded trade associations, railroad industries, and agricultural cooperatives. Over the years, it also had policies about the size of companies that were in scope, for years excluding all manufacturing companies with 50 or fewer employees. In later years, manufacturing industries assumed to have little or no R&D were excluded, but uneasiness about the extent of R&D in smaller industries led to their inclusion in later years. Variable cutoffs based on number of employees were used for many years to reduce scope. At other times cutoffs were reinstated. Single units with fewer than five employees were eliminated from scope.

Today, all for-profit companies classified in nonfarm businesses are

BOX 3-2
Changing Scope of the Industry R&D Survey

The earliest files were derived from lists of businesses reporting to the Board of Old Age and Survivor's Insurance (BOASI), which had the benefit of having industry codes classified into Standard Industrial Classification (SIC) codes. Then, for a time, the sample for the manufacturing industries was selected from the Annual Survey of Manufactures (ASM) carried out by the Bureau of the Census. The BOASI records still constituted the frame for the nonmanufacturing industries. Lists from the Department of Defense (DoD) of the largest R&D contractors supplemented both sources.

In 1967, the 1963 Census Enterprise Statistics file was the frame for multiunit manufacturing companies. Single unit manufacturers were sampled from the 1963 Economic Censuses. Social Security Administration (SSA) files represented the nonmanufacturing universe. Lists of R&D contractors supplemented the selected panel for the Department of Defense and the National Aeronautic and Space Administration (NASA). After updating to the latest economic census, the same procedure was followed for the 1971-1975 panel, with the exception that the Enterprise file was used for selected nonmanufacturing industries. The SSA files represented the remaining nonmanufacturing industries.

In 1976, a change in frame sources occurred that holds to this day. The Census Bureau's Standard Statistical Establishment List (SSEL) was used. This list is updated annually and contains all nonfarm entities that the Census Bureau knows about. Although it was still not used for some of the nonmanufacturing companies, beginning in 1981, it was the prime source for all manufacturing and nonmanufacturing companies. It is still supplemented by lists from DoD and NASA, and occasionally other sources and is now called the Business Register.

included in the coverage of the survey. There is no size criterion, except for the exclusion of single units of companies with fewer than five employees.

On a practical basis, the scope is defined and limited by the sampling frame for the survey. The sampling frame has changed over time. The frame has shifted from a social security file based on the Board of Old Age and Survivors' Insurance program, to a frame built on multiple sources, to a frame constructed using the U.S. Census Bureau's Business Register (see Box 3-2). The Census Bureau Business Register is the foundation of the Bureau's economic programs. This establishment database contains data from the Internal Revenue Service, the Social Security Administration, and the Bureau of Labor Statistics. It serves as a frame for selecting samples for all of the Census Bureau's economic programs and is updated on an ongoing basis for births and deaths.

One of the most difficult issues in the industry R&D survey is distinguishing those companies that perform R&D from among the many that do not—the problem of finding a needle in a haystack (see Box 3-3). Some of these performers are known from previous surveys or other sources, many of which are contained in the "certainty" part of the sample—firms that should be included in the survey with certainty—which now totals 1,100 units. Most of the sampled companies in the industry R&D survey perform no R&D at all. In fact, the survey imposes a burden on nine firms to find just one that does reportable R&D. However, the small number of firms that are newly discovered by the survey may report significant expenditures. Multiplying their expenditures by a large sampling weight can lead to spikes in the time series of estimated totals and inflates variances. This can be especially crucial for estimates of small populations, such as state totals.

In response to the problem of severe fluctuations in the state estimates, NSF and the Census Bureau have taken several steps. One was to develop an estimator, a composite that takes into account research on small-area estimation. A second was to consider the use of some type of smoothing across time to try and eliminate or reduce the spikes. This option is under study and has not been implemented.

A third step has been to add questions to the Company Organization Survey about the performance of R&D. The Company Organization Survey is conducted annually by the Census Bureau to obtain current organization and operating information on multiestablishment firms. The results are used to maintain the Business Register. The United States Code, Title 13, authorizes this survey and provides for mandatory responses.

The Company Organization Survey is an annual survey that collects individual establishment data for multiestablishment companies. It is designed to collect ownership and operational status information for every

BOX 3-3
Finding a Needle in a Haystack

Companies in scope for the R&D survey	1,831,849
Companies sampled in 2002	31,182
Companies with some reported R&D activity	5,808
Companies previously known to perform R&D	2,558
Companies identified as performing R&D, not already known	3,250
Discovery ratio*	(3,250:28,624 =) 1:8.8

*The ratio of the number of R&D performers found to the number of companies with unknown R&D sampled.

SOURCE: Data provided by the staff of the National Science Foundation.

establishment of a company for purposes of maintaining the Census Bureau's Business Register—the source of the sample for the industry R&D survey in 2004 (Bostic, 2003). Beginning in 2003, two questions were added to the survey to determine whether any R&D is done and, if so, the dollar amount (see Box 2-1):

- "Did this company sponsor any research and development activities during 2003?"
- "Yes—Did the expenditures on these research and development activities exceed $3 million?"

It is expected that the answers to the new questions will assist NSF and the Census Bureau in selecting a better sample for the 2004 survey. The information will permit distinguishing between companies that are R&D performers and those that are not, so that more of the sample can be directed toward companies with known R&D activity.

A fourth step is to retain the top 50 performers in each state as certainties each year. Since the noncertainty sample is independently selected each year, some performers identified in the current year may not be sampled in the next year. A rotating panel sample using permanent random numbers as described below may be a useful way of controlling this last problem.

A rotating panel sample would retain the noncertainty sample units for several years consecutively. Thus, when a sample company is found to have some R&D expenditures but not enough to make it a certainty, the company would remain in the sample for a few years and continue to contribute to the estimate instead of potentially dropping out the next year. After the system is fully under way, a portion of the sample would be rotated out every year and a replacement set of units rotated in. This is most easily implemented using stratified equal probability sampling. Currently, the noncertainty universe is partitioned between large and small companies. Large companies are sampled with probability proportional to size while small companies are selected by stratified simple random sampling. Thus, rotating panels would be straightforward to implement for the small companies while a switch to stratified simple random sampling might be needed for the stratum consisting of large companies. If the measures of size now used for large companies are not strong predictors of R&D within this stratum, then converting to equal probability sampling in strata would not sacrifice efficiency. Strata could be constructed so that uniform within-stratum rates would approximate the desired rates of probability proportional to size.

A variation on the rotating panel design is to have no overlap at all between the samples for consecutive years. This could be desirable in strata in which virtually no R&D is performed. Having 100 percent rotation would increase the possibility of picking up some performers while still maintaining a probability sample, although complete rotation of the sample would considerably weaken estimates of change.

Use of permanent random numbers (PRNs) is one simple way of implementing a rotation scheme. Ohlsson (1995) and Ernst et al. (2000) describe the PRN method and its properties. A PRN could be assigned to each company on the frame by generating a uniform random number between 0 and 1. For each stratum in the design, a sampling window would be established depending on the desired sampling rate for the stratum. For example, a sampling rate of 0.1 could be achieved by selecting all companies in a stratum with $0 < PRN \leq 0.1$. By moving the window for the next time period to, say, $0.2 < PRN \leq 0.12$ (an expected) 20 percent of the time period 1 sample would be rotated out and an additional 20 percent rotated in. This method does lead to a random initial sample size, but with a large sample size this is a relatively minor issue. (Prior to 1998, the RD-1 survey did use a type of Poisson sampling that had a random sample size.) A variant of this method, called sequential random sampling, orders the units in a stratum by PRN and selects the first n of these. At the first time period, $0.2n$ in the example above would be dropped and the next $0.2n$ in the sorted list added to the sample. Although this method will yield a fixed sample size, as long as $0.2n$ is an integer, the selection probabilities of units will change if the stratum population size changes due to births and deaths. The sampling window method more naturally accommodates births and deaths while maintaining a fixed sampling rate, as noted below. A variation on PRN sampling, called collocated sampling, uses equally spaced random numbers and can help reduce variation in sample sizes. This method is used in several establishment surveys conducted by the Bureau of Labor Statistics (Butani et al., 1998).

By assigning a PRN to new entrants to the universe and applying this sampling window method to an updated frame at each time period, the sample automatically updates itself for births and deaths. Thus, cross-sectional estimates can be made that properly represent the current universe. The amount of overlap in the initial sample can also be controlled to improve estimates of changes. Nonresponse will, of course, mean that the amount of overlap will not be totally under control, but it can be predicted, at least in an average sense.

Supplementation with Special Lists

One concern in the industry R&D survey is that small start-ups that are engaged in research and development activities are not quickly captured by the Business Register used as the frame for the survey. An effective way of including new small companies may be to use commercial lists as a supplement to the Business Register. One example is the CorpTech list of technology companies sold by OneSource Information Services, Inc. In 2004, this file consisted of 50,199 U.S. entities and 2,395 non-U.S.-owned entities, arrayed in 14,766 units. Over 50 percent of the entities in this database were in the hard-to-identify service industries, with telecommunications and Internet and computer software companies predominating. Other potential sources of data to be explored are the Dun & Bradstreet U.S. marketing file, lists of venture capital firms (Venture One, VentureXpert, and the Securities Data Company's Strategic Alliances database), and the Recombinant Capital file.

The use of dual frames is common practice when attempting to survey a universe that has certain segments that are of special interest and that may otherwise be difficult to locate (Groves and Lepkowski, 1985; Kott and Vogel, 1995). For example, in its Commercial Building Energy Consumption Survey, the Department of Energy uses several list frames of special types of buildings to ensure that the sample sizes of hospitals, schools, and large buildings are adequate (Energy Information Agency, 2001).

Finally, there is the possibility that NSF has, in its own programs, the capacity to strengthen the frame for the industrial R&D survey. The Survey of Earned Doctorates (SED) is designed to obtain data on the number and characteristics of individuals receiving research doctoral degrees from U.S. institutions. This survey collects information from over 40,000 recent graduates in science and engineering research fields, who are asked to "name the organization and geographic location where you will work or study." This information is tabulated by name of organization, geographic location, and sector (government, private, nonprofit).

Another human resource survey, the Survey of Doctorate Recipients (SDR), is designed to provide demographic and career history information about individuals with doctoral degrees and asks questions about the place and nature of work. Specifically, the survey collects information on educational history (field of degree/study, school, year of degree, etc.), the employer's main business, employer size, employment sector (academia, industry, government), geographic place of employment, and work activity (teaching, basic research, etc.). This information is available for about 40,000 individuals with research doctorate degrees. It is not a simple task to align the information provided by individuals with the Business Register information maintained by the Census Bureau; however, the advantage of being able to focus attention on the employers of people with educational credentials suggesting an R&D focus offers possibilities for identifying such employers and stratifying the frames for the purpose of estimating the amount of R&D in industry.

There are a number of practical problems to be solved in using one or more supplemental lists. Lists may overlap and duplicates must be handled in some way. The units on the lists may not all be the same—establishments may be mixed in with companies, for example—and some editing will be needed in advance of sampling. However, **the payoff in efficiency through utilization of supplemental lists could be substantial, and the panel recommends investigating this approach (Recommendation 3.2).**

Classification Issues

The industrial R&D survey is a survey of companies, not business units. The present practice of classifying R&D activity within the companies follows the standard Census Bureau practice for allocating economic activity by companies in which R&D activity is attributed to the company's *primary* industry classification. Each company is assigned a single NAICS code based on payroll. Multiunit companies are assigned a code that represents their largest activity as measured by payroll.[7]

This practice of assignment of industrial classification has evolved over the years. Initially, the classification code was that of the establishment having the largest number of employees. Later, when the Census Bureau used the annual survey of manufacturers as the sample frame, major activity for a company was based on value added, then on product shipment. Now that the Census Bureau's Business Register is the frame, the assignment is based on payroll.

As was pointed out in a previous National Research Council (NRC) report, this means, for example, if 51 percent of a firm's payroll is classified as in motor vehicles and 49 percent in other products, all of the R&D activities are classified in "motor vehicles" (National Research Council 2000:89). This lumping of R&D activity into a single code can result in overrepresentation of some major industrial activities and underrepresentation of others. The distortion is most serious in companies that are highly diversified.

In order to obtain a more accurate depiction of the industrial classification of R&D activity, it has been suggested that the data be collected by business unit within the companies. The business unit is a firm's activities associated with a given product market. If that activity could be separately identified, reported, and aggregated, data on R&D would more sharply

[7]The primary code is assigned using a three-stage procedure. The largest 2-digit economic division is determined for a company using establishment payroll. Then the 3-digit major group within the economic division with the largest company payroll is determined. Finally, the largest 4-digit industry group within the 3-digit group is determined by payroll size and assigned as the company's industry code.

depict the true nature of the R&D activity, enabling a better understanding of the character of technological innovation.

It has also been suggested that collection of data at the line-of-business level would improve understanding of the geographic focus of R&D. At present, collection of R&D data at the corporate level requests that the company submit data on R&D expenditures by state. Companies have various means of ascribing R&D activity to states. Strong anecdotal evidence suggests that companies have developed means of allocating activity by geographic area that produce data of questionable quality. If the data were collected by line of business, it might be possible to also improve the accuracy of the estimates of state R&D activity and to obtain information of substate geographical location for the first time.

For these good reasons, a 1997 NRC workshop report concluded that the industry R&D survey should be administered to business units (a firm's activities associated with a given product market) and should collect "R&D expenditures, composition of R&D (process versus product; basic research, applied research and development), share of R&D that is self-financed, supported by government, or other contract, as well as contextual information on business unit sales, domestic and foreign, and growth history of the business unit" (National Research Council, 1997:20).

Although a strong case can be made for collection of data at the business unit level, there are also key questions to be answered: Are the data available at the appropriate level of detail, are they collectable, and is the cost of obtaining the data, both in terms of collection resources and burden on the respondents, justified by the benefit?

Issues in Identifying the Line of Business

As discussed above, since the earliest days of industrial R&D classification, the practice of determining the code of a company by its main activity means, in essence, that the entire research and development operation of a company is classified in that industry (Griliches, 1980). Early on, however, there were voices that suggested the need for more detailed coding. For example, Mansfield (1980) suggested that, for some purposes, it might have been better to use finer industrial categories. Otherwise, some results for individual industry groups are difficult to interpret (Mansfield, 1980). The debate continues today. Recent work by the Conference Board has concluded that there is considerable heterogeneity in the cost structure and skills among business units of the same firm (McGuckin, 2004). These differences are masked in today's R&D data.

The problem stems from the fact that the reporting unit for the industrial R&D survey is the *company*, defined as a business organization of one or more establishments under common ownership or control. The

TABLE 3-1 Company Counts, RD-1 Data, 2000 (by company and other R&D in $million)

Sector	Total	< 0.2	≥ 0.2, < 1	≥ 1, <10	≥ 10, < 100	≥ 100
Manufacturing companies	16,917	9,805	3,777	2,559	632	145
Nonmanufacturing companies	17,456	7,781	5,260	3,443	896	76
Total companies	34,373	17,586	9,037	6,002	1,528	221

SOURCE: National Science Board (2002).

survey includes publicly traded and privately owned, nonfarm business firms in all sectors of the U.S. economy. For the latest year for which data have been released, company count data are shown in Table 3-1. The number of nonmanufacturing companies in the sample has exceeded the number of manufacturing companies, although the number of very large R&D performers in manufacturing is still larger than the number in nonmanufacturing.

Through the years, there have been a number of attempts to disaggregate the R&D data by collecting information from companies at the line of business, enterprise, or product level.

It has been noted that global companies disaggregate their data when they file the corporate income tax form with IRS, since they are obligated to show their tax liability on income for the domestic part of the company. They also distinguish R&D in the United States and abroad when they apply for the R&D tax credit. The Bureau of Economic Analysis, in its foreign direct investment surveys, requires global companies to do the same. And at the Census Bureau, when the Company Organization Survey is sent out, the target is the domestic part of the company, and the same distinction is made when enterprise statistics data are compiled and published. Thus, there appears to be ample precedent, both in government agencies and in respondent companies, for reporting R&D separately for the domestic and foreign parts of the company.

NEW DATA APPROACHES

FTC Line of Business Program

The country went through a substantial merger wave in the late 1960s and early 1970s, which was typified by the creation of some very large diversified companies. In both the financial community and the government regulatory community, concern about the loss of useful data began to grow. In the financial community, the issue was how investors could continue to make sense out of corporate data as one large company after another got acquired by the large conglomerate buyers. The result was that the Financial Accounting Standards Board issued its Statement of Financial Accounting Standard (SFAS) 14.[8]

The Bureau of Economics at the FTC was interested in standard measures of industrial organization performance variables: sales, assets, profits, advertising and other marketing expense, and R&D expense. Given specific interests of the FTC—enforcement of antitrust and consumer protection laws—the bureau designed a data collection program that would get at least the four leading companies in each industry. The target of four companies per industry was designed to provide publishable industry detail while protecting confidentiality. The FTC designed the industry list to highlight industries that were large and subject to firm and industry behavior that made them attractive from the perspective of the FTC's mandate.

The contribution of line-of-business data to understanding how companies organize and account for R&D and how R&D should be classified is exemplified in comparing the FTC results with the standard view obtained by analysis of the Census Bureau results. The purpose of the FTC effort was to disaggregate the companies' data across relatively specifically defined industries so as to be able to compute industry aggregates that would more accurately reflect industry performance. The Line of Business Program was designed to capture conventional "industrial organization" variables—a variety of profit measures, advertising expenses, and R&D expenses. Some comparative data for 1977 for the FTC and NSF/Census Bureau data collections are shown in Table 3-2. A further treatment of the history of the program is given below.

As shown in the table, 456 companies reported a total of 3,680 manufacturing lines of business, so the average number is 8.1 per company. (The true average could be less than 8.1, though probably not much less.) There could be fewer, although probably not many fewer, since each reporting company was required to report even small amounts of R&D if it had such activities.

The impact of the NSF/Census Bureau procedures of assigning a whole company to a single industry and using a relatively small number of quite aggregative industry definitions, relative to the approach taken by the FTC, can be seen clearly in Table 3-3. The ratios of company-financed R&D to

[8]SFAS 14 was replaced with SFAS 131 starting in 1998. For an empirical analysis of the impact of the change, see Street et al. (2000).

TABLE 3-2 Comparison of Federal Trade Commission (FTC) and National Science Foundation (NSF) Data, 1977

Source	Companies	Units	Units/ Company	Industries	Units/ Industry
FTC Line of Business	456	3,680	8.1	259	14.2
NSF/Census Bureau, manufacturing, all reporters	13,497	13,497	1.0	25	539.9
NSF/Census Bureau, manufacturing, all reporters, 259 industries	13,497	13,497	1.0	259	52.1
NSF/Census Bureau, manufacturing, ≥ $5 million in R&D, FTC Line of Business	507	4,107	8.1		

SOURCE: Long (2003).

sales, as well as the ranks, are taken from the 1977 FTC report. The matching ratio data for NSF are taken from or derived from their published report, and the ranks were then determined.

The first line of the table, for what the pharmaceutical industry calls ethical (prescription) drugs, is striking. The industry definition is at the finest level of detail in the FTC list, being part of a 4-digit Standard Industrial Classification (SIC) (2834 pt.). In the other column, the Census Bureau determined that, even though the pharmaceutical industry (283) is shown as one of the reporting industries, some aspect of the data required the suppression of the company-financed R&D data. The number in the table is for all chemicals except industrial chemicals (283-289). Even this is at a finer level than the one used in the Griliches study, which prompted Mansfield to observe, "[f]or example, the chemical industry includes petroleum, chemical, and drug firms. Thus, when R and D intensity and other variables are regressed on firm size, a considerable part of the relationship must be due to the well-known differences among the petroleum, chemical, and drug industries" (Mansfield, 1980:455).

Data for drugs are not the only problem, however. Third in the FTC ranking is aircraft engines and parts. The closest industry in the NSF publication is aircraft and missiles. More on the impacts of this two-pronged approach of assigning a whole company to a single industry and using very aggregative industries is presented below.

Although the FTC data were shown to be quite valuable in understand-

TABLE 3-3 Company R&D/Sales Ratios, 1977

Industry	Federal Trade Commission %	Federal Trade Commission Rank	National Science Foundation %	National Science Foundation Rank
Ethical drugs	10.2	1	3.6	8
Computing equipment	8.9	2	9.4	1-2
Aircraft engines & parts	8.4	3	2.9	10
Calculating, accounting machines	7.3	4	9.4	1-2
Photographic supplies	6.3	5	5.3	4-6
Semiconductors	6.1	6	3.0	9
Photocopying equipment	5.7	7	5.3	4-6
Optical instruments	5.5	8	5.3	4-6
Engineering and scientific instruments	5.0	9	5.4	3
Telephone, telegraph, radio, TV equipment	4.9	10	4.3	7

SOURCES: Federal Trade Commission (1985); National Science Board (2002).

ing the operations of companies, opposition arose from about a third of the expected respondents, culminating in a large lawsuit, joined by about 180 companies. Arguments by the plaintiffs were that: (1) the FTC was a law enforcement agency, so they could be sure that the data would be used only for statistical purposes; (2) the FTC did not have the legal authority to collect the data; and (3) allocation and transfer pricing practices vary across firms, so the data would be too flawed to be of use. The case went to trial, the FTC won the case, and the program was put into operation.

Data were collected for 1974-1977, but support for the program began to wane, due to business opposition and political pressure against data collection programs. The program was formally cancelled, although the FTC decided that the data already collected would be made available for research along with the microdata. The inclusion of R&D data in the report form did not in itself raise much concern during the contest over the program, because opponents apparently did not see it as particularly noteworthy.[9]

Product Group Data

To overcome some of the conceptual and collectability issues associated with assigning industrial classifications to detailed business units within companies, there have been several attempts to study R&D using R&D by product field. The product field data usually refer only to applied research and development—basic research is, by definition, not possible to ascribe to specific products and is excluded. In this framework, all pharmaceutical R&D would be counted as pharmaceuticals regardless of whether conducted by pharmaceutical companies or other industries. Early attempts to analyze product field data were limited by problems of reliability of published data (Griliches and Lichtenberg, 1984).

A recent study shows the differences in the industrial breakdown of R&D in the United Kingdom that are induced by collecting product group data rather than industry of origin of the spending. It indicates, for example, that total R&D in pharmaceuticals would be £2.5 billion if defined by product, versus just £743 thousand if defined by industry (Griffith et al., 2003).

IRI/CIMS Business Segment Reporting

Starting in 1993 and continuing through 1998, the Industrial Research Institute (IRI), working with the Center for Innovation Management Studies (CIMS), formerly at Lehigh University and now at North Carolina State University, collected R&D data from member companies for business segments. These business segments were for the most part the same as the industry segments that companies reported on their 10K filings with the SEC.

Each segment was classified into 1 of about 70 industries, most of which were at the 4-digit SIC level. For the years for which data were collected, the number of companies and business segments are shown in Table 3-4. In addition to data for business segments, the survey collected data on R&D labs. The program ceased the collection of new data with the 1998 reporting year.

---

[9]The FTC data were not without drawbacks. As pointed out in Cohen and Klepper (1992), large firms were overrepresented in the data because the FTC sample was drawn almost entirely from the largest 1,000 firms in the economy, as measured by domestic sales of manufactured products.

TABLE 3-4 Industrial Research Institute and Center for Innovation Management Studies' Business Segment Reporting

Item	1993	1994	1995	1996	1997	1998
Companies	62	83	84	79	82	77
Business segments	142	181	157	142	138	131
Segments/company	2.3	2.2	1.9	1.8	1.7	1.7

SOURCE: Bean et al. (2000).

Carnegie Mellon Survey

Under the direction of Wesley Cohen, Carnegie Mellon University collected data on a large number of manufacturing R&D laboratories, after expending a substantial amount of resources on identifying the labs and key people at them (Cohen et al., 2000, 2002a). The population sampled was all of the R&D labs or units located in the United States conducting R&D in manufacturing industries as part of a manufacturing firm. The sample was randomly drawn from the eligible labs listed in the Directory of American Research and Technology or in Standard and Poor's COMPUSTAT, stratified by 3-digit SIC industry (RR Bowker Inc., 1994).[10] The survey asked R&D unit managers questions about the "focus industry" of their unit, which the authors defined as the principal industry for which the unit conducted its R&D activities. The survey sampled 3,240 labs and received 1,478 responses, yielding an unadjusted response rate of 46 percent and an adjusted response rate of 54 percent.[11]

Cohen and associates asked R&D unit or lab managers to answer questions with reference to the "focus industry" of their R&D lab or unit, defined as the principal industry for which the unit was conducting its R&D. Thus, respondents identified the industry for which the lab conducts research.

This substantial data file has now been used to support empirical work on a number of R&D issues. The data cover only one year, but the study makes available a wide variety of variables, supporting research on the public-private question, the role of patents, the impact of spillovers, and firm size and diversity effects. Additional research based on this important study has shown that the impact of industrial R&D on university and

[10]The authors also oversampled Fortune 500 firms.

[11]The nonrespondent survey showed that 28 percent of the nonrespondents were ineligible for the survey because they either did no manufacturing or did no R&D. Adjusting the sample accordingly yielded the adjusted response rate of 54 percent of eligible respondents.

government research is much more pervasive than previously thought, suggesting that the returns to such publicly-funded research are likely greater than typically thought; proved that patents stimulate R&D across the entire manufacturing sector; and shown that patents contribute to greater R&D spillovers in Japan than in the United States, suggesting that patents can have an important effect of the diffusion of new technology.

Center for Economic Studies

Since Griliches' original work, the Census Bureau has developed the Center for Economic Studies (CES) to support the kind of research he did by pulling resources from the Census Bureau and other places together into one facility while facilitating access to outside researchers. CES now has available for use data from 1972 through 2000. A modest number of researchers have used the R&D data.

One major effort that utilized the newly available facilities of the CES has been the development of a set of master files of Census Bureau R&D survey reports with reports based on COMPUSTAT, with a special emphasis on R&D and related variables. Several dozen working papers that use these files have been released (Hall and Long, 1999).

The important work of the Center for Economic Studies highlights the potential of more intensive use of firm-level data linking R&D expenditures with other aspects of the behavior of the firm. This work suggests the need to collect and process R&D or innovation data at the firm level in such a manner that it can be integrated with the rich new longitudinal firm-level datasets that have been and are being constructed at the federal statistical agencies. Such microdata integration is essential for a number of reasons. First, such data integration permits micro-based research, which is essential for scientific analysis. Second, such data integration permits richer public domain aggregated statistics that can be constructed based on the integrated data. Third, such data integration permits a cross-check on potential data problems. For example, some of the new matched employer-employee data have rich information about the workforce composition and earnings associated with different types of workers. The latter information might be quite useful for checking against the survey responses on the number of employees devoted to R&D activities.

Can R&D Line-of-Business Data Be Collected?

It is intuitive that the business unit is the appropriate unit of collection, if it is possible to collect data at the business unit level. Many of the data on R&D activity in multiunit companies are available only at the divisional or business unit level (Hill et al., 1982). For example, to complete the RD-1 form at the company level of aggregation, as requested in the industrial R&D survey program, someone at the corporate level has to gather and collate data from business units. The panel was able to discuss this process of aggregation with one major company in which this process is formal, structured, and embedded. However, many firms do not have standardized and centralized procedures for completing these data requests.

In order to answer the central question of whether line-of-business data could be collected in the industrial R&D survey, the Special Studies Branch of the Manufacturing and Construction Division of the Census Bureau conducted a study based on a survey of 45 companies that had filed RD-1 forms for 1997 (U.S. Census Bureau, 2000). The study concluded that reporting on the RD-1 form at the line-of-business level would be inappropriate because: (1) companies use different terms to refer to lines of business; (2) companies are not able to assign the lines of business they have in industry terms; and (3) companies could not provide detail on basic research, applied research, and development.

While this study was illustrative, it was far from conclusive. Indeed, 43 of the 45 companies stated that they could report R&D at a subcompany level. The problem was a lack of uniformity as to the reportable subcompany unit. Some could report on "groups," others on "sectors," still others on "product lines." These various nomenclatures may have reflected differing organizational structures, but they could also have simply reflected naming conventions.

A critique of this survey indicated problems with analysis of nonresponse and failure to use collateral data from the Census Bureau and elsewhere (Long, 2003). Needless to say, additional study of the collectability of line-of-business data is warranted.

Revisiting Previous NRC Recommendations

This is not the first time a National Research Council study group has looked at the issue of line-of-business reporting and recommended action to NSF. After examining the appropriate unit of measure for the R&D survey, the Committee to Assess the Portfolio of the Division of Science Resource Studies of the NSF concluded that NSF should "examine the costs and benefits of administering the Survey of Industrial Research and Development at the line of business level" (National Research Council, 2000:8). We reemphasize that recommendation, and **we conclude that appropriate assignment of industrial classification to industrial R&D activity requires additional breakdowns of data at the business unit level (Recommendation 3.3).**

Since the publication of the previous NRC recommendation, NSF and the Census Bureau have taken the significant step of collecting R&D data in the Company Organization Survey. This survey collects data at the establishment level, and provides, for the first time, a platform for assessing the feasibility of collection and the attendant cost and benefit issues associated with line-of-business reporting. **We urge NSF and the Census Bureau to evaluate the results of the initial collection of R&D data in the Company Organization Survey to determine the long-term feasibility of collecting these data (Recommendation 3.4).**

NSF and the Census Bureau should test the ability to collect some disaggregated data by more detailed NAICS codes. It could take a top-down approach, as tested by the FTC, or a bottom-up approach, as utilized in the Yale and Carnegie Mellon surveys. In either case, the ability of reporters at the central office or in decentralized operating units to respond to the inquiry is the key to collecting valid line-of-business data. **We recommend that the record-keeping practice surveys should be used to assess the feasibility of respondents providing this additional detail and the burden it would actually impose on reporters. With this information in hand, NSF and its advisory committee (recommended in Chapter 8) should decide whether the collection of reliable R&D line-of-business data is feasible and, if so, for all or a subset of reporters, and at which frequency (Recommendation 3.5).** If possible, this investigation should be undertaken jointly with representatives of industry, whose cooperation will be absolutely essential to the success of the collection of these additional data.

MEASURING THE GEOGRAPHIC DISPERSION AND IMPACT OF R&D

Interest in the geographical dispersion and impact of industry R&D expenditures is keen on the part of national and local decision makers, as well as those who seek to understand the relationship between investment and outcomes. The NSF documents that support the Office of Management and Budget (OMB) approval of the industrial R&D survey claim that state statistics are among the most important and most frequently requested statistics produced from this survey. Requests for these statistics come from agencies, both public and private, in states where a great deal of industry R&D is performed and from states that are trying to spur new R&D performance.

In view of this interest, and attentive to the fact that much of the interest comes from the U.S. Congress, the development of estimates of R&D expenditures by state has been a key objective of the industry survey program for some time. It is therefore unfortunate that the estimates that are now produced are not very good. Given the current survey design and implementation, they could not be expected to be very good.

The basic problem of state estimation for the R&D survey has two sources. First, R&D is a rare event among the population of all companies.

In the survey it is reflected in the reports of only about 3,400 companies in recent survey rounds. These reports cover most but not all of the largest companies, as well as about 10 percent of the 35,000 firms that conduct some R&D in a given year. Estimates that are derived from these responses may be adequate for the United States as a whole, but they could be entirely inadequate at the state level if key R&D performers in a given state did not report. Second, some reporters may not report each year, generating estimates that indicate false growth or decline in activity.

State-level reliability requirements are taken into consideration during sample selection, but R&D estimates for states are highly volatile due to the "rare event" nature of the measured variables. The year-to-year volatility is aided and abetted by the fact that the Census Bureau selects independent samples each year for all but the certainty strata.

The response of survey managers to this problem has been threefold.

1. *Enlarge the sample.* Beginning in 2002, the sample was increased by 6,000, largely to provide for more sample units for each state.

2. *Change the sample selection process.* Also in 2002, the Census Bureau began to use the last 4 years of reported R&D survey data to assign the probability of selection. Companies that reported at least once in the 4-year period were divided into three strata: all cases were taken when reported R&D exceeded $3 million; probability proportionate to size constrained by industry and state methods were used for those reporting up to $3 million; and simple random sampling was used when no R&D was previously reported. In addition, state information from the Business Register was used for the first time to assign probabilities of selection to the 1.8 million companies for which R&D status was unknown. The top 50 firms in each state based on payroll are now selected with certainty for the sample.

3. *Improve the estimation process.* The Census Bureau uses a small-area estimation procedure with the goal of reducing the mean square error in which the estimate is originally a weighted value. The Census Bureau considered mean square error as a measure of some of the bias. To overcome this effect, the Census Bureau created a model that seeks to minimize fluctuations from year to year (see Box 3-4). The model, which does not affect variance at national level, is applied to the final data in order to smooth out the estimates. While the smoothing reduces the variability, it makes it harder for the states to analyze the meaning of the data.

The panel commends the National Science Foundation and the Census Bureau for developing this composite estimator, which takes into account research on small-area estimation. However, we recommend that additional simulations be conducted to assess the bias, variance, and mean square error of these new state estimates (Recommendation 3.6). In addi-

BOX 3-4
2001 Synthetic Estimator Used to Produce State Estimates in the NSF Survey of Industry Research and Development

$$\hat{Y}_{SynS} = \sum_{i=1}^{n} y_{si} + \sum_{I} \left[R_{IS} \sum_{i=1}^{n} (w_i - 1) y_{Ii} \right]$$

$$R_{IS} = \frac{\sum_{i=1}^{N} (1-\pi) x_{ISi}}{\sum_{i=1}^{N} (1-\pi) x_{Ii}}$$

where

y_{si} = reported R&D in state S of ith company
y_{Ii} = reported R&D in industry I of ith company
w_i = weight of ith company
x_{ISi} = payroll in industry I and state S of ith company
x_{Ii} = payroll in industry I of ith company

SOURCE: Formula provided by staff of the National Science Foundation.

tion, future research could profitably explore alternative estimators for handling outliers, drawing on the literature on finite population estimation (see, for example, Cochran, 1997; Kish, 1965).

SURVEY DESIGN ISSUES

Today's R&D survey is, in reality, two surveys in one. One survey, which has existed from the earliest days of the data collection, is a survey of the companies thought to have the largest R&D expenditures. These companies can be characterized in statistical methodology terms as a "certainty stratum."

The very first survey defined the certainty stratum as all companies within scope having 1,000 or more employees. Over the years, the criteria changed. After 1994, the size criterion based on number of employees was dropped. In 1996, the criteria were total R&D expenditures of $1 million or more based on the previous year's survey or on predetermined sampling

TABLE 3-5 Sample Size and Response Rates, 1990-2002, Industry R&D Survey

Year	Sample Size	Manufacturers	Nonmanufacturers	Response Rate
1990	1,696	about 1,400	about 300	79.3%
1991	1,648	about 1,400	about 300	85.2
1992	23,376	11,818	11,558	84.0
1993	23,923	15,018	8,905	81.8
1994	23,543	2,939	10,604	84.8
1995	23,809	7,595	16,214	85.2
1996	24,964	4,776	20,188	83.9
1997	23,327	4,655	18,672	84.7
1998	24,809	4,836	19,973	82.7
1999	24,231	4,933	19,498	83.2
2000	24,844	4,808	20,036	81.1
2001	24,809	4,505	20,604	83.2
2002	30,999	10,920	20,079	80.9

error constraints relating to individual industry estimates. The threshold was raised to $5 million in 2001 in order to reduce reporting burden on the basis that, historically, R&D costs between $1 and $5 million account for only 2 percent of total R&D expenditures.

When the changes in coverage were introduced a decade ago, they were accompanied by several other changes designed to better focus and simplify the survey. Only companies identified in the Census Bureau's Business Register as having five or more paid employees are asked to participate in the survey. Furthermore, extensive use is made of information from the Business Register to supplement reporting and reduce redundant data collection.

Despite this increase in sample size in the early 1990s, response rates stayed relatively high because, in addition to expanding the survey, the survey was bifurcated into two distinct operations, nearly unique among federal surveys (Table 3-5). A new form, the RD-1A form, was introduced to simplify collection from the expanded base of noncertainty companies. This form, which is sent to nearly all first-time reporters, collects a shorter list of information than the RD-1 form, waiving the five items asterisked below.

- Research, applied research, development
- Company-sponsored R&D expenditures in foreign countries
- R&D performed under contract by others
- Federally funded R&D by contracting agency*

- R&D costs by type of expense*
- Domestic R&D expenditures by state*
- Energy-related R&D expenditures*
- Foreign R&D expenditures by country*

The RD-1A form also makes it easy for firms that do not perform R&D to quickly report that fact. The form has a screening item for respondents to indicate that they do no R&D. Those firms that do no R&D are able to complete their reporting requirements by using a toll-free touch-tone data entry (TDE) system that consists of a series of voicemail options directing respondents according to the nature of their request. Nearly 90 percent of companies responding to the short form use the TDE system.

While revisions in survey operations have done much to enrich the coverage of the survey and reduce the burden on those companies that do little or no R&D, the ease of reporting no R&D raises a new concern: Does the relative simplicity of reporting no R&D encourage underreporting? This is an issue that could benefit from research through the program of continuous record-keeping practice surveys recommended later in this chapter.

There is also an attempt to simplify the data-gathering chore for the larger respondents. The instruction package for the longer RD-1 provides definitions and item-by-item instructions. These item-by-item instructions give methods of estimating expenditures if the company does not keep records that give exact allocations. For example, methods were given to estimate basic, applied, and development expenditures.

On the RD-1 form, companies are informed that they may report on a diskette rather than on paper. In addition, in 2004, a more extensive use of the Internet for reporting is envisioned with a new version of a web-based form being made available for respondents to use.

Despite these measures, identifying companies that conduct R&D, aside from the very largest, is a difficult matter. Companies that perform reportable R&D are rare as a proportion of all companies. The Census Bureau's sample frame is comprised of over 1.8 million companies, of which only 35,000 are estimated to be actively engaged in R&D in any given year. In the 2001 survey, forms were mailed to about 25,000 companies, and only about 3,400 of them reported that they actually conducted R&D (Bostic, 2003).

Survey managers at NSF have apparently done a good job of reducing unnecessary burden on respondents while obtaining some coverage of noncertainty firms. However, the question remains: Has enough been done to produce fully representative data? Without an independent, corroborating census of R&D, it is difficult to make a judgment as to the adequacy of survey coverage. **The panel urges NSF to continue to seek an answer to**

whether the data are representative. This will be partially achieved through analysis of the results of collection of R&D information on the Company Organization Survey, continued matches to outside administrative sources such as the SEC 10K reports, reference to patent data to identify innovating firms, comparisons with patterns and trends obtained in R&D surveys in other countries, and, under the auspices of new legislation, matching data files with surveys conducted by fellow statistical agencies (Recommendation 3.7).

Need for a Consultative Sample

During the course of its deliberations, the panel found the interaction with senior staff of a large industrial R&D performer and the membership of the Industrial Research Institute's Financial Network and the Research on Research Committee to be invaluable as sources of practical information and advice. We note that fairly elaborate consultation takes place with advisory bodies and workshops assembled in support of the federal funds and academic spending surveys. There is no such standing mechanism for the industrial R&D survey.

Prior to 1990, NSF had an active and productive Industrial Panel on Science and Technology, to which an annual inquiry was sent. This panel consisted of company officials in major R&D performing industries, including the 20 top R&D spending companies as identified by research expenditures reported on the RD-1 form. In the final survey, conducted from April through September 1989, responses were received from 72 of the 83 surveyed R&D officials. Their companies represented more than one-half of all U.S. company-funded R&D expenditures.

The primary purpose of the panel was to project the growth in R&D expenditures for the current and subsequent year, and the major factors responsible for projected changes. This projection of spending by the largest R&D companies was used to provide a timely estimate of R&D spending, given the lags in the release of results from the R&D survey. The data were benchmarked to the results of the RD-1 survey for 1988, and the estimates for the 72 respondents to the 1989 survey were then used to estimate R&D spending for industrial R&D throughout the economy.

Moreover, respondents were asked to complete special inquiries from time to time. In the final survey, for example, respondents were asked about possible government incentives, the effect of cooperative R&D ventures, and whether leading foreign-born R&D personnel were being lured back to their homelands. The results of this survey were released by NSF fairly quickly after the period of collection in a "Highlights" report.

The survey was curtailed in 1990 because of resource shortfalls and concerns over the representativeness of the panel. Over time, the inability of a panel of the largest manufacturing industry reporters to represent the totality of R&D activity in the business community became even more evident. This same shortcoming in the overall RD-1 led to the survey redesign in 1991 to be more inclusive of the range of industries performing R&D.

The panel concludes that there is value in having a regular sounding board of R&D professionals from the firms. As mentioned earlier, the RD-1 survey goes primarily to financial officials in companies who are charged with responding to government questionnaires and appropriately are responsible for keeping track of R&D expenditures. In consequence, there is a low awareness of the national R&D survey's existence among the R&D executives, who are most aware of trends in the structure and performance of R&D (McGuckin, 2004). When NSF needs information on emerging trends, concepts, definitions, and the like, it is best to go to a panel of R&D officials. Therefore, **we recommend that NSF again develop a panel of R&D experts, broadly representative of the R&D performing and R&D data-using communities, to serve as a feedback mechanism to provide advice on trends and issues of importance to maintaining the relevance of the R&D data (Recommendation 3.8).**

Microdata Analysis

The opportunity for improving analysis of the scope and impact of R&D on the economy afforded by linking R&D data with data from other collections was touched on in the earlier discussion of the impact of tax and accounting definitions. This, along with longitudinally matching responses from companies to trace the path of R&D within the firm, is a highly promising area of endeavor, and some work has been done in it. Starting in 1965, Griliches (1980) and colleagues began to develop and use the industry R&D survey data at the Census Bureau for studies of the relationship between R&D and productivity. They then linked firm-level R&D data with production establishment data in order to assess the (private) returns to R&D. This work was extended by Lichtenberg and Siegel (1991) and by Adams and Jaffe (1996). However, it has not received much attention at NSF or the Census Bureau until recently.

For some time, there has been a recognized need by the Census Bureau, the Bureau of Economic Analysis (BEA), and the National Science Board to link Census Bureau R&D data to BEA data to improve understanding of the impact of foreign investment in R&D conducted in the United States and by U.S. firms abroad. This unfilled need surfaced again in 2003 when Congress passed the Confidential Information Protection and Statistical Efficiency Act, often referred to as data sharing legislation. This law permits the Census Bureau, BEA, and Bureau of Labor Statistics to share business data for statistical purposes only. Shortly after the legislation was passed, the R&D Link Project was developed under the sponsorship of NSF.

The R&D Link Project involves linking the data from the 1997 and 1999 R&D surveys to BEA's 1997 Foreign Direct Investment in the U.S. and the 1999 U.S. Direct Investment Abroad surveys. The feasibility study match is envisioned as a win-win for all three agencies. The Census Bureau will identify unmatched companies to the BEA files that conduct research and development activities, adding them to the R&D survey sample to improve coverage. The unmatched cases are expected, at a minimum, to yield additional information on foreign companies conducting R&D in the United States and the location of R&D activities conducted by U.S. companies abroad.

BEA will augment its existing R&D-related data, identify quality issues arising from reporting differences in the BEA and Census Bureau surveys, and improve its sample frames. NSF will obtain aggregate data, which will provide a more integrated dataset on R&D performance and funding with domestic and foreign ownership detail. This new dataset is expected to lead to a more complete account of international aspects of R&D performance and funding.

If the project is deemed successful, based on the data quality, the benefits derived, and the utility of the data, it is expected to form the basis for an annual linking project. **The panel commends the three agencies for this initiative and encourages this and other opportunities to extend the usefulness of the R&D data collected by enhancing them through matching with like datasets. We urge that the data files that result from these ongoing matching operations be made available, under the protections afforded in the Census Bureau's Center for Economic Studies, for the conduct of individual research and analytical studies (Recommendation 3.9).**

Collecting Data on the Nonprofit Sector

It has been recognized for many years that a major shortfall in the collection of data from R&D performers has been the lack of coverage of the nonprofit sector. Nonprofit organizations are excluded by definition from the industry survey and are not included in the Census Bureau's frame from which the sample is drawn. Some are included in the survey of R&D expenditures at colleges and universities, but the vast majority are not covered. This failure to obtain information on nonprofit performers means that the national data are incomplete for a fairly significant (4 percent) part of national R&D performance. On an ongoing basis, NSF must estimate this important part of the total national R&D effort based on reports from federal funding sources and extrapolations of information obtained in the academic performers survey.

Collection of data from this sector is exceedingly difficult, hence NSF has mounted special efforts to collect these data just twice in the last 30 years. The first collection was in 1973; the most recent covered fiscal years (FY) 1996-1997 and was conducted over the period April 1998 to the spring 1999 (The Gallup Organization, 2000).

The difficulties arise from several factors. There is no list of nonprofit institutions, much less those involved in R&D activities, so the first task for each study must be to compile a list of potentially eligible nonprofit organizations, and that requires refining a number of list sources, each with problems of coverage and completeness. There is little information on nonprofit organizations, and what there is appears scattered among several sources, so classifying them into strata for sampling purposes is quite cumbersome and judgmental. Finally, periodic surveys of this sort often are plagued by low response rates because the collections are unfamiliar to the respondents and many fail to appreciate the usefulness of the data that result from the collection.

The final response rate for the 1996-1997 survey exemplifies these problems: only 40 percent (adjusted for out-of-business and nontraceable nonrespondents) of the sample responded. There was considerable confusion over the questions and, as a result, a large item nonresponse. Over 45 percent of the questionnaires of R&D performers required some type of data cleaning or hot-deck imputation. The low response rate means that the analytical possibilities for the data set were severely limited. NSF could not publish state-level estimates, for example.

Nonetheless, there were several important lessons learned in the conduct of the FY 1996-1997 survey that can instruct a future effort of this kind:

- Infrequent surveys require long periods of rather expensive development and pretesting of the questionnaire and field procedures. The cost of this rather large investment should be included in the cost estimate for the survey.
- The process of developing the questionnaire should include involvement of focus groups and cognitive interviews.
- Much attention should be paid to developing, refining, and unduplicating the frame for the survey.
- The survey operators should focus on obtaining buy-in from survey respondents and organizations representing groups of respondents prior to the survey and devote sufficient resources to follow-up activities.

Despite these very evident problems that were well documented in the methodology report on this survey, **the panel recommends that another attempt should be made to make a survey-based, independent estimate of the amount of R&D performed in the nonprofit sector (Recommendation 3.10).** Taking into account the lessons learned in the FY 1996-1997 survey, NSF should devote resources to laying out the design for a new survey utilizing the input of modern cognitive science and sampling theory. If possible, the resurrected survey should be planned from the beginning as a continuing survey operation in order to build a stable frame, and to introduce certainty strata and other survey efficiency and quality improvement features that are not possible with periodic surveys.

Statistical Methodology Issues

Of several challenging issues in statistical methodology related to the survey, the major problem is that, following a major revision in the early 1990s, changes to the survey in the past decade have been piecemeal and incremental, impeded by the lack of resources to modernize the survey operations. Today's industrial R&D survey stands in contrast to the other surveys in the NSF portfolio, in that the statistical methodologies and technologies employed in the survey are far from cutting-edge.

Survey Design

The basic sampling frame is the Business Register, previously known as the Standard Statistical Establishment List. This list, used since 1976, has problems in coverage and currency. The Business Register may have particular problems with completeness of coverage; this potential undercoverage may be an important issue for measurement of R&D expenditures by type or by state. In view of the growing recognition that small firms are increasing as a share of research and development spending, a problem of undercoverage of small firms may well lead to a growing problem of underestimation (Acs and Audretsch, 2003).

The sampling procedure has also evolved, while maintaining a constant focus on companies with the largest R&D expenditures (1,700 companies in 2000), with additional strata defined by industry classification and various expenditure-level cutoff schemes. The largest firms were first identified by the number of employees, and later by the total R&D expenditures based on the previous year's survey. The industry strata are defined as manufacturing, nonmanufacturing, and unclassified, with 48 categories represented in the 2002 round.

To improve state estimates, take advantage of historical data, and improve industry-level estimates, the sample was further divided in 2001 to represent "knowns" and "unknowns." The known segment is comprised of companies that reported R&D expenditures at least once in the previous three survey cycles. These various sampling changes, each introduced to achieve a worthwhile objective, have made it very difficult to achieve control over sampling error in the survey estimates. The variability of the estimates has been a constant concern in the survey. Some of the sampling errors, as computed, remain very high, particularly for the nonmanufacturing universe. High sample errors and high imputation rates for some of the key companies mean that the quality of the published data are suspect, suggesting the need for evaluation of the statistical underpinnings of the survey.

Data Collection

The method of data collection currently relies on two forms: the RD-1, sent to known, large R&D performers, and the more limited RD-1A, sent to small R&D performers and companies that have not previously reported. Both surveys collect sales or receipts, total employment, employment of scientists and engineers, R&D expenditures information, character of the R&D (basic research, applied research, development), R&D expenditures in other countries, and R&D performed under contract by others. In addition, the RD-1 forms collect information on federally funded R&D by contracting agency, R&D costs by type of expense, domestic R&D expenditures by state, energy-related R&D expenditures, and foreign R&D expenditures by country.

Although a diskette is provided and a web-based version of the form is available for completion and mail-in, the primary mode of data collection is by postal mail. Survey forms are mailed in March, with a requested completion date of 30 or 60 days later. Mail follow-up is fairly extensive, with a total of five mailings to delinquent Form RD-1A recipients, but telephone follow-up is limited by resource constraints to the 300 largest R&D performers. If no response was received and no current-year data reported, several data items are imputed. The data collection procedures employed by the Census Bureau for the industry survey stand in stark contrast to the more technologically advanced procedures employed in the smaller federal and academic surveys, which, for the most part, are web-based and have more intensive education, response control, and follow-up schemes.

The printed questionnaire is in dire need of review and substantial revision. The last full-scale review of the cognitive aspects of the industry R&D survey was reported in 1995 by Davis and DeMaio (U.S. Census Bureau, 1995). This internal Census Bureau study identified a number of possible improvements in graphics and question wording in the questionnaire. Some of the graphical suggestions have been implemented, but a key recommendation that all survey items be put into question format has not yet been acted on. The wide range of suggestions for possible question wording changes were based on the level of respondent understanding of the concepts and definitions implied in the questions.

The issues developed in U.S. Census Bureau (1995) strongly support the need for an ongoing dialogue between the data collectors and data providers in the industry R&D survey. An active program of respondent contacts and record-keeping practice surveys that would have supported a dialogue was dropped in the early 1990s because of resource constraints. Such a program need not be expensive or overly intensive. Davis and DeMaio were able to obtain valuable insights into cognitive issues with just 11 company visits and a mail-out study of some 75 companies that incorporated cognitively oriented debriefing (COD) questions.

There are opportunities for improving respondent contacts within the ongoing program. For example, the Census Bureau now conducts an intensive nonresponse follow-up program for the 300 largest R&D performers. Perhaps more useful information could be gleaned from these telephone contacts, which are designed to secure cooperation or to clarify information, if a cadre of telephone interviewers trained in cognitive aspects of telephone interviewing probed more deeply into areas that would improve the data. Supplementing the regular follow-up program could yield invaluable data on the firms with little cost or additional burden on the reporters.

The panel strongly recommends that the National Science Foundation and the Census Bureau resume a program of field observation staff visits to a sampling of reporters to examine record-keeping practices and conduct research on how respondents fill out the forms (Recommendation 3.11). In this recommendation, the panel adds its voice in support of the OMB directive, which gave approval of the industry R&D survey in 2002 with the proviso that "a record-keeping study should be done to find out what information businesses have regarding the voluntary items and the reasons for nonresponse to those items" (U.S. Office of Management and Budget, 2001).

The problem of nonresponse, however, does not apply solely to items collected on a voluntary basis. As with all surveys, some sample units do not respond at all, or they omit some items. The response rate from 1999 to 2001 was in the 83 to 85 percent range. When there is additional attention, as is the case with the largest 300 companies, response rates can be higher. The rate for these large companies was 90 percent in 2000. The impact of new rules requiring mandatory reporting in the 2003 survey cycle is not yet known.

Item nonresponse is also a problem, for both the five mandatory data items and the voluntary items. It is difficult to assess the scope and impact of the problem of nonresponse, since no item nonresponse rates are given for any items. Instead, as mentioned earlier, the Census Bureau publishes imputation rates, which can be quite large and serve as a poor proxy for item nonresponse rates because the imputation rates are weighted to proportion of the total contributed by imputation. In a skewed distribution, they will be very different from nonresponse rates. NSF could make a significant contribution to understanding the quality of the data by ensuring clear definitions and regular reporting of item nonresponse rates. Furthermore, the recent rise in the proxy measure—imputation rates—gives cause for concern and impetus to the need to understand the quality impact of item nonresponse.

Survey Content

Closely associated with issues of data collection and the cognitive aspects of survey design are issues of data content and related questions of usefulness to potential data users. One problematic aspect of the survey content is the questionable validity of data that break down R&D into the components of basic research, applied research, and development. This issue is discussed earlier in this chapter.

A second content problem relates to estimates of the number of R&D scientists and engineers employed by the company, which derive from the definition: scientists and engineers with a 4-year degree or equivalent in the physical and life sciences, mathematics, or engineering. Aside from the obvious problem of transferring the form between the financial office of the company, where the form is completed, and the personnel department, where personnel records are kept, the definition is in some aspects quite vague. What does "equivalent" to a 4-year degree mean? Even if it is possible to identify field of educational specialization, what is a person's field of employment in the company? When an employee works on multiple tasks, how should they be apportioned?

Identification of the location of R&D activity by state is also problematic. Although there is intense interest in the location of R&D activity, there is anecdotal evidence that respondents have great difficulty in accurately allocating this activity to geographic areas, and consequently they have developed measures for geographic allocation that produce data of questionable quality. NSF needs to conduct a more intensive study to determine the quality of state breakdown of R&D activity and to implement changes if warranted.

Processing

Quality problems can crop up in a survey at the stages of data entry and editing. According to a series of memoranda by Douglas Bond that discuss research on various sources of processing errors (U.S. Census Bureau, 1994), data entry procedures produced little error, but the editing process was replete with potential for error. That potential starts with the observation that there is no written description of the editing process, including the process in which an analyst supplies data codes. This study is now a decade old. Although many of the sources of editing error may have been corrected, the absence of a more recent study of processing error and the lack of current documentation of the editing process cause concern over the impact of this source of error.

The panel recommends that the indudtrial R&D editing system be redesigned so that the current problems of undocumented analyst judgment and other sources of potential error can be better understood and addressed (Recommendation 3.12). This redesign should be initiated as soon as possible, but it could later proceed in conjunction with the design of a web-based survey instrument and processing system.

Imputation is an integral part of the survey operation, and the rates of imputation are high. Again, Bond's studies found several sources of potential error in an imputation process that varies with the item being imputed. There should be an ability to clearly determine whether errors arise in the editing or the imputation processes.

MANAGEMENT STRATEGIES

The recommendations concerning the industrial R&D survey are the panel's highest priorities. There is an urgent need for the survey to be better managed. This can be achieved in a number of ways, including:

1. Finding a contact person to whom the survey is sent for each respondent.
2. Assigning a person at the Census Bureau or NSF to each contact person to answer questions and discuss aspects of the survey.
3. Creating a standing committee of contact people or high-level R&D people who could discuss all issues pertaining to the survey and who could be queried about the usefulness of potential changes in the survey.
4. Increasing NSF involvement in the administration or implementation of the survey, whereby NSF more closely oversees the work done for it by the Census Bureau.
5. Reporting and publication of the R&D data in a more timely manner.

The panel discusses the industrial R&D survey and makes recommendations for redesign in Chapter 8.

4

Measuring Innovation in Business and Industry

Innovation encompasses but is more than research and development. It is becoming apparent that much technological innovation does not result from traditional research and development, particularly in the service industries (Guellec and Pattinson, 2000). The main activities involved in innovation are, in addition to R&D, other acquisition of knowledge (patents, licenses, and technical services); acquisition of machinery and equipment (both incorporating new technology and for standard use when producing a new product); various other preparations for production and delivery, including tooling up and staff training; and internal and external marketing aimed at the introduction of the innovation (Organisation for Economic Co-operation and Development, 1992). Although many of these activities extend well beyond the subject of this report, nonetheless it is useful to discuss the measurement of innovation because of the close interaction between traditional R&D and the process of innovation.

Innovation measures must cover five activities: the introduction to the market of new products; the development of new processes to produce, or deliver, products for the market; the development of new markets; the finding of new sources of supply of raw materials; and changes in the organization of firms. Introducing new products to the market has implications for economic growth, and new processes provide opportunities for improvements in productivity, quality, or other desired objectives, such as reduced environmental emissions or a happier labor force.

Moreover, there is a general feeling that the role of technological innovation in improving productivity and increasing economic growth is increasing, as is international economic competition for domestic and international markets. Therefore, it is important to understand the innovation process and why some companies, industries, and countries are apparently more innovative and enjoy a greater gross domestic product (GDP) per capita than others.

The panel observed that one reason for the strong interest in measuring innovation activity in Canada is that the GDP per capita for Canada is $28,000 (United States; adjusted for purchasing power[1]) but in the United States it is $36,000 US. This disparity is in part attributed to an evident "productivity gap" that is not being resolved by the market. This, in turn, has given rise to policy interventions to close the gap. In Canada, this takes the form of a 10-year innovation strategy, which aims to make Canada one of the world's most innovative countries. European countries face a similar gap. Emphasis on innovation policy in Europe was affirmed at a special meeting of the European Council held in Lisbon in 2000, which set a strategic goal for the coming decade of making the European Union "the most competitive and dynamic knowledge-based economy in the world, capable of sustainable economic growth with more and better jobs and greater social cohesion" (European Council, 2000). The "Lisbon strategy" was reaffirmed and strengthened by the European Commission in March 2003 (European Commission, 2003a). Developing countries in Latin America, including Argentina, Chile, and Mexico, among others, have also conducted innovation surveys.

If this perceived shortfall in productivity attributed to a relative lack of innovation is the driving force for interest in innovation studies abroad, is there a similar rationale for the gathering of information on innovation by NSF in the United States? Although interest in systematic innovation measurement has been less intense in the United States, it is widely recognized that innovation information is valuable for offering insights about best practices and where they can be applied with greatest effect. It is also recognized that there are intraregional and structural dimensions to innovation in the United States. The GDP per capita of the state of Georgia, for example, is less than that of California, and presumably less than that of several European countries, and the propensity to innovate in pharmaceuticals is greater than that in wood products industries.

Although economic statistics can identify a productivity gap, understanding why the gap exists requires information about firms. This can

---

[1]Purchasing power parities (PPPs) are the rate of currency conversion that eliminates the differences in price levels between countries. They are used to compare the volume of GDP in different countries. PPPs are obtained by evaluating the costs of a basket of goods and services among countries for all components of GDP (Organisation for Economic Co-operation and Development, 2002b).

come from case studies, surveys, or a combination of both. Surveys that identify the introduction of new or significantly changed products or processes by firms can relate the activity of innovation to the broader economic context that led the firm to innovate. They can also identify the barriers to innovation, the sources of information and technologies used to innovate, and the impacts of the activity, such as a change in the level of employment in the firm or in the skill levels required by the workforce as a result of the change. All of this can be related to productivity measures.

SURVEYS OF INNOVATION IN INDUSTRIALIZED COUNTRIES

Innovation surveys are not as well developed as those for industrial research and development expenditures. R&D surveys emerged after World War II and their concepts and definitions were eventually codified by the Organisation for Economic Co-operation and Development (OECD) in 1963 with the first *Frascati Manual* (Organisation for Economic Co-operation and Development, 1963).

Innovation surveys have gone through a long experimental period in many industrialized countries. Hansen and Hill (Hill et al., 1982, 1983) led this work in the United States in the 1980s. There were surveys in the Nordic countries at the same time, as well as in Canada and Germany, where the first annual survey of innovation was put in place. Some of these early surveys were quite ambitious. In his review of technology innovation surveys, Hansen (2001) identified the largest effort as one conducted in Italy in the 1980s (see Table 4-1 for major innovation surveys, 1979-2001). The Italian National Research Council and Central Statistical Office sent a survey to the 35,000 firms in Italy with more than 20 employees, which revealed that many more firms were engaged in innovation activities than just those conducting in-house R&D. They followed up with a detailed questionnaire to the 25,000 firms that reported some innovation activity in response to the first survey, which asked about number and types of innovations, types of underlying technologies, sources of information, obstacles to and costs of innovation, and impact on sales (Archibugi et al., 1991; Cesaratto et al., 1991).

These initiatives were rather uncoordinated and experimental, so concepts and definitions differed and comparison was difficult. As with the *Frascati Manual*, innovation surveys came into their own only when the OECD provided a forum for developing a common set of concepts and definitions codified in the first *Oslo Manual* (Organisation for Economic Co-operation and Development, 1992).

The *Oslo Manual* was used to guide the first Community Innovation Survey (CIS.1) in the European Community in 1992. The manual dealt only with technological innovation and with manufacturing. Experience

TABLE 4-1 Surveys of Innovation at the Firm Level, 1979-2001

Survey	Year	Country	Responses	Method	Focus
Ifo Institute	1979	Germany			Sources of innovation
Italian National Research Council-Institute of Information Science and Technologies	1986	Italy	24,700 firms	Two-stage survey	Sources of innovation
Yale	1982	United States	650 respondents 42% response rate	R&D manager survey	Appropriability role of science
French: 1. Piatier 2. Observatory of Science and Technology	Late 1980s	France	1. 5,300 respondents 2. 15,000 respondents	Questionnaires	Sources of innovation
Nordic Innovation Surveys	1989	Nordic countries	650 respondents	Questionnaires	Sources of innovation
Community Innovation Survey 1 (CIS-1)	1992	12 European Union countries and Norway	Circa 40,000	Differences in questionnaires	Sources of innovation
Canadian Innovation Survey 1	1993	Canada	*Oslo Manual* 5,000 responses	Questionnaire	Sources of innovation
Policies, Appropriation, and Competitiveness in Europe (PACE) Survey	1993	United Kingdom, Ireland, Benelux, Italy, Germany, France, Denmark	R&D managers 706/1,200 return	Yale type questionnaires	Appropriability role of science

Carnegie Mellon Survey 1 and 2	1994	1. United States 2. Japan	1. 1,478 (54% response rate) 2. 643 (53% response rate)	Yale II - same questionnaire	Appropriability role of science
Canadian Innovation Survey 2	1996	Canada	5,000 responses 100% return	Questionnaire	Sources of innovation
Community Innovation Survey 2 (CIS-2)	1997	15 European Union countries, Norway, and Iceland	16,950 manufacturing firms (38% response rate overall—median: 65%) 11,932 service enterprises (7% response rate)	Questionnaire	Sources of innovation
Canadian Innovation Survey 3	1999	Canada	5,700 95% return	Computer assisted telephone interviewing	Sources of innovation
Community Innovation Survey 3 (CIS-3)	2001	15 European Union countries, Norway, and Iceland	Not yet released	Questionnaire	Sources of innovation

SOURCE: Adapted from Science and Technology Policy Research (expanded and revised) (http://www.sussex.ac.uk/Users/prff0/RM1/Autum_2002/SurveyII%20week%20VIII%202002.doc).

gained from surveys led to the first revision of the manual, jointly with the statistical office of the European Union, Eurostat. The second edition (Organisation of Economic Co-operation and Development and Eurostat, 1997) still dealt with technological innovation, but its industry coverage was expanded to include all of the private sector. It guided CIS.2, conducted in 1997, and CIS.3 in 2001. By 2003 it was clear that the manual needed further revision to take account of nontechnological innovation, such as the adoption of new business practices and changes in organization of the firm. The revision was launched by the OECD and Eurostat in June 2003, with completion anticipated in 2005.[2]

Not all surveys of technological change have focused on the introduction of new products or processes. In the United States, for example, there have been surveys that concentrated on the ability of firms to capture, or appropriate, the results of their inventions and their interactions with universities and national laboratories (Levin et al., 1987). The Yale I survey, as it became known, was conducted in 1982 and was aimed at high-level R&D managers who had knowledge of both the technological and market conditions facing their firms. This was followed by the so-called Carnegie Mellon survey in 1994 (Cohen et al., 2000, 2002a). The Carnegie Mellon survey built on but went well beyond the Yale I survey by collecting measures on the source and channels of knowledge affecting industrial R&D, the regional sources of that knowledge, a large number of measures of R&D activity, the patenting behavior of firms, the actual uses of patents, the intensity of technological rivalry, the impacts of public research on industrial R&D, the actual uses of patents, the management of innovation in the firm, and R&D performance. This design was intended to capture measures of the determinants of industrial R&D and R&D performance.

[2]The *Oslo Manual* provides definitions and methodologies for collecting data on corporate strategies, the role of diffusion, sources of innovative ideas and obstacles to innovation, inputs to innovation, the role of public policy in industrial innovation, the outputs of innovation, and the impacts of innovation (Archibugi and Sirilli, 2001). The categories of information to be collected were: prevalence of the innovation (number of firms, industries); types of innovation (products, processes); goals of innovation (improved or radically new product, new market, higher quality, better performance); internal sources of innovation (in-house R&D, sales and marketing, management); external sources of innovation (suppliers, clients, university or government laboratories, technical literature); practices for protecting innovations (patents, trademarks, trade secrets, complexity of industrial design); intensity of innovation (ongoing, occasional); obstacles to innovation (lack of skilled personnel, high risk, lack of information, regulatory barriers); impact on workers (number of employees, productivity, skills); and impact on economic performance (percentage of sales attributable to new or improved products or processes) (Hansen, 2001).

Cohen was also central to a collaboration with the National Institute of Science and Technology Policy in Japan that resulted in surveys of R&D directors and laboratory managers across major innovation-intensive industries in the United States and in Japan and Europe in 1994.[3] The purpose of the survey was to improve the understanding of the factors affecting innovative activity across nations in order to provide a more informed basis for designing national and multilateral policies that promote technological progress leading to economic growth (Cohen et al., 2002b).

In Europe, the Policies, Appropriability and Competitiveness for European Enterprises (PACE) project included a survey of managers of the 500 largest manufacturing firms in the European Union (Arundel et al., 1995). The survey collected data on the goals of innovation, external sources of knowledge, methods to protect intellectual property, and legal or regulatory impediments to innovation. The survey was similar to the Yale I and II surveys.

While there was an interest in how knowledge was acquired and protected, there was also an interest in what technologies were being used in industry and how they were diffusing. This led to surveys conducted by the U.S. Census Bureau, the first of which was done in 1988 (U.S. Census Bureau, 1989). Similar surveys were done in Australia and Canada (Ducharme and Gault, 1992), and, as a result, the first edition of the *Oslo Manual* included a section on the importance of measuring the use and planned use of manufacturing technologies, which remained in the second edition (Organisation for Economic Co-operation and Development and Eurostat, 1997).

As with the case for the industry R&D survey, there is much to learn by turning attention to initiatives north of the border. Statistics Canada has conducted several surveys of innovation, beginning in 1993, to better understand innovation in Canada:

- The 1993 Survey of Innovation and Advanced Technology, which surveyed manufacturing firms.
- The 1996 Survey of Innovation, which surveyed the communications, financial services, and technical business services industries.
- The 1999 Survey of Innovation, Advanced Technologies and Practices in the Construction and Related Industries, which was the first survey of advanced technologies and practices in the construction sector.
- The 1999 Survey of Innovation surveyed manufacturing and was the first innovation survey of industries in the natural resource sector.

---

[3]See www.cgp.org/cgplink/vol04/researchvol04html.

The surveys were designed in accordance with the guidelines contained in the *Oslo Manual*, although they were not the same as (harmonized with) the CIS surveys. In fact, for various reasons, the Canadian survey findings cannot be compared completely with the results of CIS-2 (see Mohnen and Thierrien, 2003, and Therrien and Mohnen, 2003, who compare Canada's Survey of Innovation 1999 with CIS-2 data from France, Germany, Ireland, and Spain). For one thing, because it is the country's only statistical agency and because completion of its surveys is a legal requirement, Statistics Canada is able to field fairly long surveys and also link easily with its other datasets (for example, its industrial R&D survey data). This means it can collect a broader range of data and go into greater detail than the CIS surveys. For example, the Canadian survey asks about employee development and training programs, employee access to the Internet, and use of the Internet by the firm to sell its products. It collects much more detail on a firm's objectives in undertaking innovation, the firm's sources of knowledge relevant to innovation, the barriers to innovation, and methods of protecting the results of innovation (these examples are all from Hansen, 2001).

The experience in Australia is also instructive. The Australian Bureau of Statistics (ABS) conducted two innovation surveys in the 1990s using the conceptual framework of the *Oslo Manual*. A third was to be conducted in 2004. Interest in innovation during this period shifted from a singular focus on measurement issues, to a broader interest in understanding the roots of innovation. In March 2000, ABS announced it was moving away from the idea of directly measuring innovation. "Instead, we are investigating whether we can produce a range of statistical indicators on the knowledge-based economy and society." In 2001, the prime minister launched "Backing Australia's Ability: An Innovation Action Plan for the Future."

ABS had already begun developing statistics to measure a knowledge-based economy and society and released a framework for discussion in 2002. The discussion paper proposed approximately 125 statistical indicators grouped in broad dimensions to enable assessment of the degree to which Australia is a knowledge-based economy and society. The dimensions are innovation, and entrepreneurship; human capital; information, and communications technology; context; and economic and social impacts. Within each dimension there are a number of proposed characteristics, and within each characteristic are potential statistical indicators that provide quantitative measures of that characteristic. For instance, one proposed characteristic of the context dimension is social and cultural factors; indicators for this characteristic might include the age structure of the population, the income level and distribution of the population, and participation in community activities (Australia Bureau of Statistics, 2002).

INNOVATION AND NSF

NSF was involved in work related to innovation in industry in the 1960s (National Science Foundation, 1967; Myers and Marquis, 1969). In 1967, the Charpie Report called attention to the inadequacies of government policies and management practices concerning innovation (U.S. Department of Commerce, 1967), which resulted in the establishment of the National R&D Assessment program at NSF in 1972. During the 1970s, the program supported research on the impact of industrial management practices and government policies on industrial innovation.

Another period of interest began in the 1970s, in response to concern that the United States was losing its technological edge, that there was little productivity growth, that government regulations were slowing innovation, and that industry was not doing enough basic research. The White House responded by conducting a Domestic Policy Review of Industrial Innovation in 1978, which was followed by the President's Commission on Industrial Competitiveness in 1983 and the creation of the Council on Competitiveness in 1986. During this period, NSF supported special efforts to develop indicators and a survey of industrial innovation conducted at the Massachusetts Institute of Technology and Boston University.

NSF's interest in innovation has slowly evolved from experimental studies to piloting official surveys. In 1994 NSF and the Census Bureau conducted a pilot survey of 1,000 respondents in manufacturing and one service-sector industry (computer programming, data processing, and other computer-related services) that followed the *Oslo Manual* (National Science Foundation, 1996). The survey found that innovation was fairly widespread, with one-third of firms reporting recent introduction of new products and processes or plans to introduce a new product or process in the near future. Not surprisingly, computer hardware, precision instruments, pharmaceuticals, and chemicals firms reported levels of innovation well above the average. Although the survey yielded interesting results, it suffered from a relatively low response rate (57 percent), underscoring the difficulty in collecting this kind of innovation information. The problem of low response has plagued innovation surveys both here and abroad over their developmental cycle.[4]

In 2003, NSF conducted a second innovation survey. This time the focus was on the information technology sector and a sample of companies

---

[4]The preliminary response rates for countries in the CIS.3 round of surveys ranged from 22 to 80 percent (Larsson, 2004).

outside information technology that were extensive purchasers of information technology products and services. Again, the concepts and definitions were drawn from the *Oslo Manual.* The final response rate for this survey was again about 57 percent (2,005 of 3,504 firms completed the survey). The results are now being analyzed.

LESSONS LEARNED

What has been learned from more than 20 years of measuring innovation in many countries is, first, that innovation can be measured, along with its linkages and outcomes and its economic and social environment. However, it is necessary to learn the second lesson before the information resulting from the many measurements can be widely used.

The second lesson is that common concepts and definitions are necessary to provide guidance to those conducting the surveys and interpreting the results. The forum for the development of these concepts, definitions, and guidelines for the interpretation of the results was provided by the OECD in the late 1980s, leading to the first edition of the *Oslo Manual* in 1992.

The *Oslo Manual* guided the first European Community Innovation Survey (CIS.1) and other surveys and provided boundaries and direction for the work—boundaries that survey teams began to challenge from the moment the manual was published. The third lesson is that the formal structure of the manual and the ongoing measurement activity combine to provide a dynamic learning environment for survey practitioners and users of the new information.

Fourth, measurements that can be made, codified, and developed in a learning environment can be used. The results of innovation surveys influenced policy development in many OECD countries and especially in Europe (European Commission, 2003b).

Other lessons have been learned about the feasibility of measuring the means used by firms to acquire and protect intellectual property in the United States, Japan, and Europe and to make comparisons across these countries (Arundel, 2003).

In summary, **the panel concludes that innovation, linked activities, and outcomes can be measured (Conclusion 4.1).** They are measured successfully in other countries, and, except for survey management issues here in the United States, have been shown to produce potentially useful measures of U.S. innovative activities. At a basic level, information on innovation in U.S. firms could provide contextual material for the work of the Economic Development Administration as it guides "innovation-led development" (Economic Development Administration, 2001).

While this may suggest support for a program of innovation measurement and analysis, there are still lessons being learned (Gault, 2003) that argue for a more cautious approach. These lessons deal with the period of observation (which is currently 3 years in the *Oslo Manual* but is clearly industry dependent) and the appropriate unit of observation (Are some questions better put at the level of the firm, or should they go to the plant or the establishment?). The entry level for the novelty of innovation in the *Oslo Manual* is "new to the firm," which gives quite high rates of innovation in some industries. An additional issue has to do with the measurement of world-first and market-first innovations, which are more interesting from the competitiveness perspective. Finally, there is the question of nontechnological innovation and how that should be measured.

RECOMMENDATIONS

The need to understand the process of innovation is of critical importance to answering key questions about the source of growth of the U.S. economy. There have been several successful efforts in other countries to develop such measures, and, although lessons are still being learned, there is a growing science to support innovation measurement. The Science Resources Statistics Division (SRS) has dipped into these waters twice in the past decade, with mixed success. The division needs resources and the capacity to explore the impact of innovation on the U.S. economy. This can be done by commissioning surveys and analyzing and publishing the results, as well as by supporting academic research.

After the analytical capacity has developed in SRS and its network of experts has been established, the SRS may wish to propose, based on its findings, a more comprehensive set of measures of technological change comparable to those that now exist for research and development.

The panel recommends that resources be provided to SRS to build an internal capacity to resolve the methodological issues related to collecting innovation-related data. The panel recommends that this collection be integrated with or supplemental to the Survey of Industrial Research and Development. We also encourage SRS to work with experts in universities and public institutions who have expertise in a broad spectrum of related issues. In some cases, it may be judicious to commission case studies. In all instances, SRS is strongly encouraged to support the analysis and publication of the findings (Recommendation 4.1).

The panel recommends that SRS, within a reasonable amount of time after receiving the resources, should initiate a regular and comprehensive program of measurement and research related to innovation (Recommendation 4.2).

5

Measuring R&D Spending in the Federal Government

The question of how much the federal government is spending on research and development should be answerable in a fairly straight forward manner. One would hope for a tally that includes all federal government agencies that engage in research and development; that there would be a consistent definition of R&D applied across the government; and that the question could be answered in a standard manner with recurrent collections associated with continuing processes that would minimize the burden of collecting the information and enhance data quality. However, none of these expectations are met in today's Survey of Federal Funds for Research and Development. The survey does not have a comprehensive list of all activities; it is not consistently defined and applied across the government; and it is not associated with ongoing budget and financial management processes in many agencies. Most importantly, it is not completed in a timely manner, so its worth is somewhat diluted by the age of the data when they finally become available.

It is important to get this survey right. It is widely considered to be a primary source of information about federal funding for R&D in the United States by government, academia, and the science community. Data from the survey are expected to take their place alongside the U.S. Office of Management and Budget's (OMB's) budget authority information that underscores decisions on the federal Science and Engineering (S&E) budget and serve as the basis for the extensive program of analysis of the federal S&E budget by the American Association for the Advancement of Science and other organizations. The possible shortcomings in the survey are highlighted in recent studies that have examined the source of the discrepancy between reports of the level of federal spending for R&D as reported in the federal funds survey and the amounts that performers in industry and academia report as receiving from the federal government. (These issues are addressed in Chapter 7). When data collections fail to fully meet the needs of users, substitutes spring up. As might be expected, there is a growing, new competition to the survey in the form of a microdata-based collection known as the RaDiUS database, also funded by NSF.

In this chapter, we focus mainly on the current status of the survey of federal funds and its deployment within the federal government. In view of the uses put to the data as addressed in Chapter 2, we examine the adequacy and relevance of this data collection. Finally, we consider the RaDiUS database to determine the extent to which this data source compliments or competes with the federal funds survey. We recommend a reconsideration of several aspects of this survey operation to modernize it and improve its operations. Attention is also paid to examining these issues as they pertain to the annual, congressionally mandated survey of federal obligations to academic institutions.

- Is the accounting framework appropriate, that is, the use of obligations rather than authorizations or outlays?
- Can federal agency staffs that are asked each year to provide information about their R&D obligations be motivated to regard this task as a benefit rather than a burden?
- Are the federal funds collections maximizing the ability to report data from agency accounting systems in the form requested by NSF? Can the collection be placed on a microdata level, that is, comparable to Level 5 in RaDiUS?

FEDERAL FUNDS SURVEY

The federal funds survey collects data on federal support of national scientific activities in terms of budget obligations and outlays. For each year, survey data are to be provided for three fiscal years (FY)—the FY just past, the current FY, and the president's budget year. Actual data are collected for the year just completed, while estimates are obtained for the current year and the budget year.

Agencies are asked to submit the survey data on the same basis as the budget authority figures submitted to the OMB in January. Definitions of obligations and outlays are purposely the same as those in the U.S. budget—*obligations* represent the amount for orders placed, contracts awarded, services received, and similar transactions during a given period, regardless of when the funds were appropriated and when future payment of money is required, and *outlays* represent the amounts of checks issued and cash payments made during a given period, regardless of when the funds were appropriated or when the obligations were incurred. The reporting unit for the survey is the subagency or agency to whom the survey materials are sent.

The survey is sponsored and funded by the National Science Foundation (NSF) and is carried out under contract by QRC Macro.

The scope of the survey has changed over the years, reflecting the number of subagencies that now fund R&D, the kinds of places in which R&D is conducted, and the kinds of questions agencies have the resources to answer. Federal obligations for research to universities and colleges by agency and detailed science and engineering field were added to the survey in 1973. Federally Funded Research and Development Centers (FFRDCs) are also included. The Central Intelligence Agency and other security-related agencies are not included.

In order to reduce respondent burden, this survey is tailored to the respondents:

- All are asked for information on total obligations and outlays for R&D and R&D plant; obligations for basic, applied, and total research, by field of S&E and by performer; and obligations to R&D plant, by performer.
- Some data are collected only for the immediate past year. These include obligations to individual FFRDCs and to foreign performers.
- Only the 10 largest R&D funding agencies are asked about the geographical distribution of obligations for R&D and the equipment and facilities where the R&D takes place. (These agencies account for about 97 percent of total R&D and R&D plant obligations each year.)
- Six agencies are asked to report on the distribution of R&D to universities by field of science and engineering and character of research.
- Only the Department of Defense (DoD) is asked to break out state obligations by research and development separately.

Furthermore, NSF has further been sensitive to the ability of federal agencies to provide good data and has removed items, as well as added items, over the years. Among the data removed from collection are data for the special foreign currency program and detailed field of S&E data for estimated out-years. Some data were removed but later reinstated by popular demand. Data on foreign performers by region, country, and agency are an example of this reconsideration.

To align with the president's budget process, the frame for the survey is the list of federal agencies that report funding R&D obtained from information in the president's budget submitted to Congress. In FY 2003, the 29 agencies and 73 subagencies that have reported R&D data in OMB budget documents comprised the census for the data collection.

The collection process focuses on key respondents in the agencies. For the most part, the persons assigned by agencies to complete these forms are budget analysts, not experts in the conduct of R&D. In this way, they are much like the normal respondents in the industry survey, who tend to be in the comptroller and accounting end of corporate operations. The respondents are identified by the contractor, QRC Macro, which keeps a master list of all respondents by agency and subagency, updating that list throughout the survey process. Quality and timeliness begin with these respondents. They are counted on to know their agencies, develop some expertise in concepts and definitions, doggedly followup to obtain responses from across the agency, summarize a variety of responses into an agency total, and report to the NSF contractor via the Internet.

This process renews itself about the same time each year, when packets are mailed and respondents are granted access to the system to begin data entry. This is usually around the first week of March. The agencies are asked to return the completed questionnaires by April 15. Few are able to meet this deadline. The responses and subsequent processing drag on to the extent that data on federal obligations and outlays with a reference period of September 2001 were not available in brief form until June 2003, or in full detail until April of 2004—some 2½ years later. (Issues of timeliness for all of the NSF surveys are addressed in Chapter 8.)

The discussion of measurement problems in the survey reports is limited. Some agencies find it difficult to report certain items. Some agencies, such as the Department of Defense, do not include some relatively minor headquarters planning and administering of R&D programs in the full cost of R&D. R&D plant data are also underreported because of difficulties encountered by some agencies, particularly DoD and the National Aeronautics and Space Agency (NASA), in identifying, compiling, and reporting these data.

It is not surprising that federal agencies should report difficulties in reporting accurate and consistent data over time to NSF. In a presentation to the 1998 NSF workshop on federal R&D data, Robert Touhy, director of program analysis in the Office of the Director, Defense Research and Engineering, stated that the collection of the data is intensive and difficult to administer, and that the benefits barely justify the high cost to DoD in labor and resources (Touhy, 1998). He identified antiquated software, inflexible collections, and the fact that DoD divisions frequently take "shortcuts" as problems those compiling the data face; he recommended no less than a reengineering of the process of federal funds data collection as the solution to these problems.

The reporting of federal funds is governed by common concepts and definitions but anchored in agency contract management and classifications systems that may generate results that are at variance with the overall definitions. Interviews conducted for a previous National Research Council study determined that agency respondents were working "with contract classification systems that evolved long ago and may no longer fit current classifications of fields of science and engineering" (National Research Council, 2000:95).

Some agencies report difficulties in distinguishing activities between the categories of basic research, applied research, and development. In the 1998 NSF workshop, it was pointed out that some research may be fundamental (basic) but have a strong relevance to the agency mission, so it could be classified as applied (National Science Foundation, 1998). In some agencies, however, the classification systems are much more robust, creating a rich analytical database that is unfortunately lost to that purpose when forced into the three basic classifications of R&D. Such is the case with the DoD science and technology (S&T) financial reporting structure. The financial reporting framework breaks S&T spending into nine classifications, most of which illuminate "development" but these distinctions are lost when aggregated into the three standard classifications of R&D (see Table 5-1).

Accounting Framework for Federal Funds

Over the years, there have been several efforts to supplement the federal budget classifications that underpin the federal funds survey with more practical or illustrative classification structures. One such proposal, which has been partially adopted, is the federal science and technology (FS&T) budget framework recommended in *Allocating Federal Funds for Science and Technology* (National Research Council, 1995). This framework would focus on investment in the creation of new knowledge and technologies and exclude activities not involving the creation of new knowledge or technologies, such as the testing and evaluation of new weapons systems. Specifically, the FS&T budget would exclude nongeneric technology development at DoD and the Department of Energy (DoE).

The FS&T budget was reexamined in a 2002 National Research Council report and was adopted by OMB in the annual budget process. It has provided an annual tabulation of the results in the "Analytical Perspectives: Research and Development" section of the president's budget, explicitly labeling the presentation as the "Federal Science and Technology Budget" beginning with the FY 2002 budget proposal (U.S. Office of Management and Budget, 2004). While it is certainly useful and feasible to present the budget in this format, the classification requires numerous judgments about activities to include and exclude. The FS&T budget also includes all costs associated with these programs, including staff salaries, in contrast to the federal funds data. Other features include the incorporation of key science and engineering education programs at the National Science Foundation that are not considered R&D but are critical investments in science and technology, as well as the presentation of identifiable line items in the budget to permit easy tracking through the Congressional appropriations process (National Research Council, 2002).

Although it is generally agreed that the FS&T budget is useful as a tool to analyze the federal budget, there is less agreement that the federal funds survey should collect and publish an FS&T obligations (and outlays) series. The inability by users to extract the entire FS&T budget detail from the federal funds survey was reported as a disadvantage of the current provision of FS&T in the 1998 federal funds workshop (National Science Foundation, 1998). To generate such a series would require the collection of more explicit and disaggregated reports by activity than are currently collected by NSF. In order to replicate the FS&T budget authority information published by OMB, the federal funds survey would need to collect programmatic information not now collected that would extend the data the collection beyond the normal R&D, particularly in the civilian agencies. Nonetheless, to meet this demand, NSF has collected breakouts of DoD major weapons systems developments from DoD for the past four years and has recently begun breaking out atomic energy defense program spending at DOE.

In view of the important uses of the federal science and technology budget, **the panel recommends that NSF continue to collect those additional data items that are readily available in the defense agencies and expand collection of expenditures for those activities in the civilian agencies that would permit users to construct data series on FS&T expenditures in the same manner as the FS&T presentation in the president's budget documentation (Recommendation 5.1).**

Enhancing Information About the Federal R&D Portfolio

In view of the differences in interpretation and application of the concepts and definitions of R&D among the agencies, as well as the growing desire to publish different "cuts" of the data, such as the FS&T reporting of obligations and outlays, the panel considered whether the federal funds survey data could be collected and preserved at a highly disaggregated (microdata) level with detail at the contract, project, program, and activity levels. However, an attempt to do just that in the construction of the RaDiUS database by the Science and Technology Policy Institute of RAND has identified complexities of the task that need to be taken into account

TABLE 5-1 Definitions of R&D

	Government-Wide	
	OMB Circular No. A-11 (1998)	
Conduct of R&D[b]	Basic Research	Systematic study directed toward greater knowledge or understanding of the fundamental aspects of phenomena and or observable facts without specific applications towards processes or products in mind.
	Applied Research	Systematic study to gain knowledge or understanding necessary to determine the means by which a recognized and specific need may be met.
	Development	Systematic application of knowledge toward the production of useful materials, devices, and systems or methods, including design, development, and improvement of prototypes and new processes to meet specific requirements.

	DOD-Unique[a]	
	DOD Financial Management Regulation (Volume 2B, Chapter 5)	
S&T Activities[c]	Basic Research (6.1)	Systematic study directed toward greater knowledge or understanding of the fundamental aspects of phenomena and or observable facts without specific applications towards processes or products in mind.
	Applied Research (6.2)	Systematic study to gain knowledge or understanding necessary to determine the means by which a recognized and specific need may be met.
	Advanced Technology Development (6.3)	Includes all efforts that have moved into the development and integration of hardware for field experiments and tests.
	Demonstration and Validation (6.4)	Includes all efforts necessary to evaluate integrated technologies in as realistic an operating environment as possible to assess the performance or cost reduction potential of advanced technology.
	Engineering and Manufacturing Development (6.5)	Includes those projects in engineering and manufacturing development for service use but which have not received approval for full-rate production.
	Research, Development, Test and Evaluation (RDT&E) Management Support (6.6)	Includes R&D efforts directed toward support of installations or operations required for general R&D use. Included would be test ranges, military construction, maintenance support of laboratories, operations and maintenance of test aircraft and ships, and studies and analyses in support of R&D programs.

continued

TABLE 5-1 Continued

	Government-Wide
	OMB Circular No. A-11 (1998)
R&D Equipment	The acquisition of major equipment for R&D. Includes expendable or moveable equipment (e.g., spectrometers, microscopes) and office furniture and equipment. Routine purchases of ordinary office equipment or furniture and fixtures are normally excluded.
R&D Facilities	The construction and rehabilitation of R&D facilities. Includes the acquisition, design, and construction of, or major repairs or alterations to all physical facilities for use in R&D activities. Facilities include land, buildings, and fixed capital equipment, regardless of whether the facilities are to be used by the government or by a private organization, and regardless of where title to the property may rest. Includes such fixed facilities as reactors, wind tunnels, and particle reactors. Excludes movable R&D equipment.

[a]Does not pertain to the Corps of Engineers.

[b]Includes administative expenses. Excludes routine product testing, quality control, mapping, collection of general-purpose statistics, experimental production, routine monitoring and evaluation of an operational program, and the training of scientific and technical personnel.

	DOD-Unique[a]	
	DOD Financial Management Regulation (Volume 2B, Chapter 5)	
	Operational System Development (6.7)	Includes those development projects in support of development acquisition programs or upgrades still in engineering and manufacturing development, but which have received Defense Acquisition Board (DAB) or other approval for production, or production funds have been included in the DoD budget submission for the budget or subsequent fiscal year.
	Developmental Test & Evaluation	Efforts associated with engineering or support activities to determine the acceptability of a system, subsystem, or component.
	Operational Test & Evaluation	Efforts associated with engineering or support activities to determine the acceptability of a system, subsystem, or component.
No separate definition	Major equipment dollars are mixed with the dollars for the "Conduct of R&D" and carried in the RDT&E accounts (i.e., 6.1 through 6.7) listed above. In FY 1998, DoD requested a total of $68M for major R&D equipment.	
No separate definition	In FY 1998, close to 90 percent of the $67M requested by DoD for R&D facilities was carried separately in Military Construction accounts. The rest were included in the costs of major development programs and are mixed with the dollars for the "Conduct of R&D"[c] carried in the RDT&T accounts (i.e., 6.1 throgh 6.7) listed above.	

[c]Includes costs of laboratory personnel, either in-house or contractor operated.
SOURCE: Fossum et al. (2000:615).

should the federal funds survey be conducted based on disaggregated data. Indeed, the ability to distinguish R&D activities from science and technology activities of the federal government was described as the biggest challenge to creating the RaDiUS database (Fossum et al., 2000). This pioneering database development project, sponsored by the National Science Foundation, illuminates us on the promise and pitfalls of building depth into the view of federal R&D investments.

The RaDiUS database is designed to connect the aggregated data on authority, obligations, and outlays with disaggregated data on actual spending (purchases) to complete the picture of federal R&D activities. With the exception of funds spent on the construction and rehabilitation of federal R&D facilities and the purchase of major R&D equipment, the RaDiUS database explicitly utilizes the published federal information for the top 24 agencies that spend R&D dollars. The contribution of the RaDiUS project is that it then disaggregates the data by subagencies and, finally, into the records, of actual obligational transactions to their final destination at universities, laboratories, and centers. The database contains over 500,000 records, which are said to provide details on close to 80 percent of the activities in the federal R&D portfolio.

The database is constructed largely by harvesting data from various transactional administrative data sources throughout the federal government. These sources usually exist for other purposes and were not designed with the objective of identifying R&D activities with precision.[1] For each transactional record culled from these databases, it is necessary for the RaDiUS builders to construct a record that depicts the hierarchial levels that permits aggregation back into agency and subagency totals. Depending on the data source, the detailed records contain (in addition to agency-identifying information) identifying numbers, type of funding, and estimated start/end dates; performer, performer type, and performer location; contact information, place of performance, funding information, and an abstract. These records are updated on schedules specific to each input source. For example, DoD records are updated 2-3 times a year, and NASA and DoE records are updated annually.

---

[1]Among the sources included in the RaDiUS data files are the Catalogue of Federal Domestic Assistance; the U.S. Department of Agriculture's Current Research Information System; the U.S. Department of Health and Human Service's Computer Retrieval of Information on Scientific Projects and the Information for Management, Planning, Analysis, and Coordination system; the DoD's R-1 and R-2 Budget Exhibits and Technical Effort and Management System; the DoE's Laboratory Information System; the Federal Assistance Awards Data System; the Federal Procurement Data System; the OMB's MAX System; the Department of Veterans Affairs' R&D Information System; the NSF's Science and Technology System (STIS); and NASA's RAMIS System.

The RaDiUS database is an ambitious undertaking, and it has been utilized to support analysis of sectoral and geographic investment in R&D. Most recently, the database served as the source of information for a report on federal research and development funds provided to colleges and universities (Fossum et al., 2004). The database is not immune from the vagaries of federal administrative classification systems, however. RAND reports that federal procurement officials decide which activities are defined as R&D on a daily basis, with reference to federal procurement regulations. The dividing line between R&D and "studies and analysis" is particularly blurry, and it is possible that there is some misclassification of activities at their procurement source that would affect the data.

As sponsor of the RaDiUS project, NSF has conducted several external reviews of the system to provide an independent review of the database. In a 1998 staff review, NSF concluded that the R&D project database was only 65 percent complete, with large parts of the DoD and DoE development investment portfolio not accounted for (National Science Foundation, 1999). The coverage was estimated by RAND at "close to 80 percent of the activities in the R&D portfolio" in its 2000 report (Fossum et al., 2000:613). Likewise, the quality of some of the input data is uneven. It was noted that some agencies, such as the National Institutes of Health (NIH) and NSF, provide very polished and comprehensive project descriptions, while other agencies provide project titles or minimal descriptions.

The passage of the E-Government Act of 2002 has added a new sense of urgency to considering the proper level at which to tally federal R&D activities. The act calls on NSF, in conjunction with OMB and the Office of Science and Technology Policy, to create a database and web site to track federally funded research and development (U.S. Office of Management and Budget, 2003b).

The panel is encouraged by the initiative to provide detailed and additive information on federal R&D activities that is represented by the RaDiUS database. Although it has several drawbacks, including the inability to extract important field-of-science data from the administrative data sources, the use of administrative records represents a valuable addition to the stock of information on the federal R&D portfolio. The committee does not suggest that an administrative database will ever supplant the collection of federal funds data on a regular basis. For example, RaDiUS data is not available by field of science. However, the RaDiUS-type data can improve the timeliness of the federal statistics, and they may allow less frequent collection of the greater detail in the federal funds survey. The promise and possibilities of a microdatabase on federal R&D activities will not be fully realized until the heterogeneous data formats and R&D procurement practices of the several key federal R&D agencies are brought into consistency.

The panel urges NSF, under the auspices of the E-Government Act of 2002, to begin to work with the Office of Management and Budget to

develop guidance for standardizing the development and dissemination of R&D project data as part of an upgraded administrative records-based data system (Recommendation 5.2).

Statistical Methodology Issues

The Survey of Federal Funds for Research and Development is a virtual census of federal agencies that support national scientific activities. It consists of collection of some data that have rigid definition and strict accounting measures. However, others, such as current and future year obligations, are best estimates on the part of agency reporters.

The trigger for inclusion of an agency in the annual data collection round is reporting of R&D activities in OMB budget documents or in the media. Rather intense effort is given to identifying respondents in the selected agencies and contacting respondents at multiple intervals—before, during, and after data collection. This is not a full-time job for respondents, and many of them are replaced from year to year, so reeducation is a constant challenge. Data collection is tiered, in that larger R&D agencies are asked to provide data on obligations to states and to colleges and universities by field of science. The response rate for the census is an enviable 100 percent, and there is no item nonresponse, since the agency respondents must answer all questions before the data can be submitted.

Lack of timeliness is a continuing issue. In his remarks at the panel's July 2003 workshop, Kei Koizumi pointed out that data for FY 2001, which ended on September 30, 2001, were just becoming available in July 2003, over a year and a half later. This delay in assembly and transmittal of the data is particularly troublesome because reporting offices have the data at the end of the fiscal year (interview with Mark Herbst, Office of the Secretary of Defense, March 26, 2003). The lack of timeliness is such a severe deterrent to utility of the data that most organizations that assess R&D spending trends turn to budget authority data rather than obligations and outlays, limiting timely application of some very useful R&D analysis tools, such as the Federal Science and Technology Budget. Although several steps are being taken by the contractor to enhance cooperation and speed data processing, the list of problems inhibiting on-time reporting is a constant challenge to the staff. Speaking at an agency workshop, Ronald Meeks of NSF identified as timeliness issues the lack of support from senior officials in some agencies, the need for constant reeducation of reporters, problems with completion of the automated form, delays from internal review and controls, and timing conflicts with the higher priority president's budget (Quantum Research Corporation, 1999).

SURVEY OF FEDERAL SCIENCE AND ENGINEERING SUPPORT TO UNIVERSITIES, COLLEGES, AND NONPROFIT INSTITUTIONS

NSF has considerable experience in collecting microdata from federal government agencies, primarily from contract award and grants files, and has done so annually since 1965. It also has experience in gathering data that are more encompassing of the scope of science and engineering than the more restrictive traditional R&D definitions that underscore the federal funds survey. These innovations in data collection are part of the annual Survey of Federal Science and Engineering Support to Universities, Colleges, and Nonprofit Institutions. This survey is, in a manner of speaking, a precursor for the kind of data collection that could significantly enrich the understanding of the nature of the federal government's investment in science and engineering.

Supplement to the Federal Funds Survey

As it stands, this congressionally mandated survey is a useful supplement to the federal funds survey. It is the only source of comprehensive data on federal S&E support to individual academic and nonprofit institutions. Federal policy makers, state and local government officials, university policy analysts, R&D managers, and nonprofit institution administrators use it to assess S&E investment trends. NSF and other federal agencies also use it for internal administrative purposes.

The federal support survey asks agency representative responders from the 21 federal agencies that incur nearly all of the obligations for federal academic R&D to rake through their contract and grant files to compile several pieces of information for each academic institution and each nonprofit institution for which they have records. The agencies are asked to compile, for each institution, the following information:

- Academic institution
- Geographic location (within the United States)
- Highest degree granted
- Historically black colleges and universities
- Obligations
- Performer (type of organization doing work)
- R&D plant
- Type of academic institution (historically black and others)
- Type of activity (e.g., R&D, S&E instructional facilities)
- Type of institutional control (public or private)

Individual contract and grant record data are aggregated to institution totals, since just one survey instrument is to be filled out for each of the over

1,000 universities and colleges for which an agency obligated R&D funding during the previous fiscal year. When reports are gathered from several subagencies in large R&D organizations like NIH and DoD, the reports for each institution must be aggregated to funding agency totals, coded, and entered into the NSF collection formats.[2] Needless to say, this is a cumbersome and time-consuming process even when agencies have automated systems that compile the basic information.[3] In order to align with the OMB budget information, the data are to be collected on a fiscal year basis (October 1 through September 30), with collection to start in the following February and to be submitted by April 15.

NSF has made extensive changes in recent years to streamline the collection, submission, and editing pieces of the survey operation. This was one of the first data collections converted to web-based systems and automated editing procedures. The contractor, QRC Macro, has an extensive program of outreach and education for the agency reporters. Despite the attempts to streamline the processes over which NSF and its contractor have some control, the burdensome nature of this collection is reflected in the lack of timeliness. The data for the most recent available fiscal year, 2002, were not published in full detail until 18 months after the reference period.

Data are to be provided in fairly extensive categories:

- Research and development
- Fellowships, traineeships, and training grants
- R&D plant (facilities and equipment)
- Facilities and equipment for instruction in S&E
- General support of S&E
- Other activities related to S&E
- All other activities

Aside from the category of "research and development" for which a definition is well codified and fairly widely understood, there is the possibil-

---

[2]Agency reporters are responsible for finding and entering a 6-digit Federal Interagency Committee on Education (FICE) code on each record. This requires reference to a 400-page Code Book for Compatible Statistical Reporting of Federal Science and Engineering Support to Universities, Colleges, and Nonprofit Institutions. This process may lead to coding errors, and perhaps some loss of coverage, particularly among nonprofit institutions, when the institutions do not get added to the list of "new" institutions in the code book.

[3]The automated systems may ease data collection, but also they may lead to errors in the process of identifying and tabulating the individual record data. Changes in software have been known to have an influence on the level of federal support reported. For example, NSF was forced to cease publication of non-S&E support estimates in FY 1993 when the Department of Education revamped codes in a major software modification.

ity that rather ambiguous definitions of the other categories, particularly those pertaining to non-R&D S&E, may lead to agency difficulties in matching program descriptions to the proper funding category. For example, NASA has placed increased emphasis on including educational components to projects, and education is always reported as "other S&E." NSF warns data users that categories of "general support for S&E" and "other S&E activities" are catchall categories (National Science Foundation, 2001).

Data on over 1,000 nonprofit institutions, each to be submitted on a separate form, are also collected, but they cover only the activities of research and development and R&D plant. Thus it is not possible to develop an estimate of S&E spending for nonprofit institutions.

The variables in this survey use definitions comparable to those used by OMB and the federal funds survey. Respondents are told that the totals reported in this survey and the federal funds survey for R&D and R&D plant obligations should be in close agreement. If differences exist, respondents should include an explanation.

Not every federal agency is in the survey every year, so, for some, there is no opportunity to build experience in compiling the data from year to year. The number of agencies included in the survey can vary from year to year depending on their activity reflected in the prior year's federal funds survey for academic and nonprofit data, so some may not have anything to report in the year of interest. Since not all federal agencies are surveyed, some funding could be missed. The omissions are believed to have little impact on understanding the total funding picture, but NSF warns users that the impact could be significant for understanding the funding for some institutions.

The recent publication of estimates of federal R&D spending in universities and colleges based on the RaDiUS database provides a glimpse at the potential for utilizing the individual record files from contracts and grants for identifying federal support for the colleges and universities, excluding FFRDCs (Fossum et al., 2004). The RaDiUS database is enriched with information from sources that are secondary to agency sources: the Federal Procurement Data System, which tracks all federal contracts; the Federal Assistance Awards Data System, operated by the Census Bureau which tracks federal grants and cooperative agreements; and the Catalogue of Federal Domestic Assistance, compiled by the General Services Administration, of all programs, projects, services, and other types of assistance provided to nonfederal entities.

Again, the approach taken by the RaDiUS database project to aggregate national R&D totals from individual record data have several shortcomings in terms of coverage and comprehensiveness, representing, as they do, a view of R&D limited by the coverage of the input data sources. Nonprofit institutions are not covered. Nonetheless, the publication of the extensive detail on universities and colleges for FY 1996 to FY 2002, in a favorably timely manner compared with the publication of NSF's federal support data, suggests the potential of data systems based on individual records of contracts, grants, and cooperative agreements to produce useful estimates.

The panel recommends that NSF devote attention to further researching the issues involved with converting the federal support survey into a system that aggregates microdata records taken from standardized, automated reporting systems in the key federal agencies that provide federal support to academic and nonprofit institutions. In connection with this investigation, NSF should evaluate the possibility of collecting for nonprofit organizations the same science and engineering variables that pertain to academia (Recommendation 5.3).

Statistical Methodology Issues

There are several difficult aspects of the survey methodology. For example, the frame for the survey is the list of all federal agencies that sponsor R&D, obtained from the president's budget submission to Congress. In practical terms, the survey covers respondents to the federal funds survey, focusing on only the largest of the agencies (in FY 2000, 18 agencies were in the target population). Unlike the federal funds survey, this survey is not a census of science and engineering support. Not all agencies are surveyed, and some funding can be missed. While the overall amount of missed funding is not significant, the patterns of funding by agency and recipient may be somewhat distorted by these omissions.

The data collection is web-based, with automated functions supporting all data collection, data imports, data editing, and trend checks. There is no nonresponse from agencies and no item nonresponse, since the forms must be completed prior to transmittal, raising issues that need to be studied, as discussed before. It is possible that coding errors, such as an incorrect institutional code or incorrect branch of a multiunit institution, could lead to errors in the estimates of funding by institution. Matching program descriptions to proper funding categories may cause some confusion on the part of respondents.

6

Measuring R&D Activity in Academic Institutions

The third component of the nation's information on R&D expendi tures is the data on expenditures in colleges and universities, broadly defined to include Federally Funded Research and Development Centers (FFRDCs). Together with funding information provided by governments and spending information provided by business and industry, the academic surveys contribute to the compilation of the extent and character of R&D spending.

SURVEY OF RESEARCH AND DEVELOPMENT EXPENDITURES AT UNIVERSITIES AND COLLEGES

The Survey of Research and Development Expenditures at Universities and Colleges serves as the primary source of information on separately budgeted R&D expenditures in academia in the United States. The survey encompasses a large part of the R&D enterprise and a very important sector: universities account for about 13 percent of total R&D performed in the United States, and FFRDCs add another 6 percent of total spending.

Importantly, the survey covers a sector with many unique characteristics that both define areas of interest and delimit the nature of the data collected. For example, most of the R&D spending by colleges and universities comes from outside sources, primarily governments and industry, so it is important to follow the funding trails in order to understand the nature of the R&D activity in this sector. Likewise, a good part of R&D is associated with the educational mission of the institutions, so it is important to consider funding in light of educational offerings and their outputs—usually expressed in the form of degrees granted but often in works cited or other indicators of academic output.

It is generally acknowledged that the survey of research and development spending at colleges and universities is a successful data collection program. NSF points with pride to the fact that the response rate from academia has historically been about 95 percent, and it is usually 100 percent for the FFRDCs (see Table 6-1). However, the high unit response rate is tempered by bothersome high item nonresponse rates for some data items. Both the small unit nonresponse and the higher item nonresponse are masked by the fact that NSF imputes values when there is no response. The validity of the imputation methods is open to question.

The survey has impressive depth of coverage: it covers all institutions with doctoral programs in science and engineering (S&E) fields or at least $150,000 in separately budgeted R&D activity, meaning that it is a virtual census of all research and development spending at colleges and universities. These institutions have traditionally expended more than 95 percent of U.S. academic R&D funds. In addition, the survey population includes all FFRDCs that are academically administered and engaged in basic or applied research, development, or management of R&D activities. Also, all historically black colleges and universities (HBCUs) that perform any separately budgeted R&D in S&E are included.

The development of the frame for the survey of colleges and universities is a very complicated undertaking, with the possibility of generating coverage error (see Box 6-1). The application of the dollar limit to identify universities and colleges that conduct R&D could be a problem in that, if the frame fails to identify an academic institution with at least $150,000 in separately budgeted R&D, there would be a gap. The HBCUs are well identified so do not pose any coverage loss. Similarly, FFRDCs are well known. The doctorate-granting universities are well known. So it is the application of the dollar limit that could pose problems. NSF describes the only other gap as the four 2-year degree-granting institutions that accounted for less than 0.01 of the total R&D expenditures. The issues raised with this frame also affect the Survey of Science and Engineering Research Facilities, which uses this frame as its universe.

While there are other sources of information on the extent and direction of R&D activity at colleges and universities, including the NSF sister surveys of science and engineering graduate students and postdoctorate students, there is no survey that provides such a long-term (annually since 1972) view of science and engineering activity at the nation's academic institutions. The historical continuity of the data is often considered to be a particular strength of this data collection, in view of the fact that the results of this survey are often blended with the results of the federal funds and

TABLE 6-1 Changes in Sampling and Coverage of the Academic Survey

Year	Total	Doctoral	Master's	Bachelor's	Response Rate	FFRDC	Response Rate (%)
1990 s	460	328	76	56	98.3%	18	100
1991 s	459	327	76	56	97.3	19	100
1992 s	46	330	77	54	99.0	19	100
1993 p	681	350	202	129	96.9	19	100
1994 s	500	349	84	67	99.6	18	100
1995 s	499	348	83	68	90.3	18	88.9
1996 s	493	343	84	66	97.3	18	94.4
1997 s	493	343	84	66	98.0	18	100
1998 p	556	358	118	80	98.6	17	100
1999 p	597	359	148	90	98.5	17	100
2000 p	623	362	162	99	97.3	16	100
2001 p	609	359	158	92	95.2	36	100
2002 p	636	353	153	94	96.2	36	100

NOTE: In some years, a sample was selected, in others, a census was conducted; s = sample, p = population.

industry surveys into a time series to portray the totality of and trends in R&D activity. However, although the survey is often presented as a continuous time series since 1972, when the survey became an annual collection after the initial four biennial collections, in reality the series has been subject to several perturbations (see Box 6-2).

In fiscal year (FY) 1978, a different population was sampled and different questions were asked, so data are not comparable for that year.

Survey design has also shifted over time:

- After 1993, the sample was truncated to include all doctorate-granting institutions but only a partial sample of nondoctorate-granting institutions.
- Significantly, current year and recent years' data are only provisional. As described by NSF, when the review of data for prior years reveals discrepancies, the prior years' data are sometimes modified.
- Prior to 1997, there was potential double-counting of R&D spending because funds that were passed through the universities to subrecipients were not broken out from the funds that were actually spent at the institution. This problem was addressed in years subsequent to 1998 when questions were added to the survey to identify "passed through R&D expenditures to subrecipients." Now, a new problem is posed for the survey, in that the level of the series adjusted for double-counting is *higher* than the level of the series without double-counting.

BOX 6-1
Development of the Survey Frame

The frame for the survey is developed from a variety of sources. Not all academic institutions are included, so a procedure has been developed for the NSF-National Institutes of Health Survey of Graduate Students and Postdoctorates in Science and Engineering (graduate student survey) and the NSF Survey of zFederal Science and Engineering Support to Universities, Colleges and Nonprofit Institutions populations. Institutions with a highest degree-granting status of 1 (doctorate-granting) are compared to make sure all S&E doctoral degree-granting institutions are included in the academic R&D expenditures survey population. This task is done annually. Sources consulted are the *Higher Education Directory* and direct contact with universities. NSF maintains a list of all FFRDCs. The Department of Education maintains a list of all HBCUs.

Each S&E doctoral degree-granting institution, FFRDC, and HBCU that reported zero expenditures in the previous fiscal year is contacted to find out if it has expenditures data to report for the current year. If there are expenditures to report, a survey packet is sent. If the institution says it has zero expenditures, a zero is entered for that institution. If the institution says zero but the federal R&D obligations data suggest otherwise, this discrepancy is discussed with the institution, and an attempt is made to obtain actual expenditures.

S&E master's and bachelor's degree-granting institutions that reported $150,000 or more in S&E R&D expenditures in the previous fiscal year are surveyed again. To determine which other institutions should be included in the current FY survey, ORC Macro creates a list of all other S&E master's and bachelors' degree-granting institutions in the academic R&D expenditures survey universe that received federal obligations during any cycle from the previous 5 years. Institutions are included in the current cycle if they either: reported cumulative expenditures for the last three full population surveys of $250,000 or more or were reported by federal agencies to have cumulative obligations of more than $400,000 for the fiscal year plus the 4 previous years.

A total of 36 academically administered FFRDCs were included; 4 FFRDCs administered by industrial organizations and 16 administered by nonprofit organizations were surveyed but not included in publications.

Relevance

In contrast to some of the other NSF data collections, there is evidence that the data are extensively used by the respondent institutions that provide the data. The data are published at the institutional level of detail, a feature of this survey that is fairly unusual among government surveys, in which the confidentiality of responses is closely protected. As a result, universities and colleges are able to use the published data from the survey to benchmark their programs against other institutions. They are also able to assess the mix and direction of their programs in the context of the

BOX 6-2
Development of the Academic Survey

Before 1998 Until 1998, full population surveys were conducted about every 5 years. During intervening years, a sample of institutions was surveyed. For academic institutions, the population consisted of all universities and colleges that offer a doctorate degree in science and engineering (certainty stratum), all historically black colleges and universities and university-system campuses (certainty stratum), and all others that reported separately budgeted S&E R&D expenditures of $50,000 or more.

1998 Beginning with the FY 1998 survey, a census of the population is surveyed annually. All university-administered FFRDCs are surveyed annually. Only S&E bachelor's and higher degree-granting institutions are surveyed. (Previously 2-year degree-granting institutions could be surveyed.) The population includes institutions with a minimum level of R&D expenditures of $150,000 or more. The threshold was raised from $50,000 to account for inflation.

2001 In order to address confidentiality suppression by the Census Bureau, an abbreviated academic survey form was sent to all FFRDCs (including 4 industry-administered FFRDCs and 16 nonprofit FFRDCs, not just university-administered FFRDCs). For the first time, the survey notice and mail-out were entirely electronic, including use of e-mail and the Internet.

general population because of the rich variety of indicators collected and published. There has been an attempt to update the data items collected to accommodate analysis of trends, primarily through refinement of definitions, inclusions, and collection of optional data items (see Box 6-3).

Together with the NSF-NIH Survey of Graduate Students and Postdoctorates in Science and Engineering and the Survey of Science and Engineering Research Facilities, the survey gives a picture of scope and depth of an institution's science and engineering program. In order to accomplish this, the survey collects the following data items on a recurring basis:

- Academic institution
- Character of work (basic research, applied research, or development)
- Equipment expenditures
- Expenditures for S&E R&D
- Federally funded R&D centers
- Field of S&E
- Geographic location (within the United States)

BOX 6-3
Modifications and Clarifications to the Academic Survey

1990 Metallurgical and materials added as a sixth engineering subdiscipline.
1991 Optional item on indirect costs/cost-sharing.
1992 Optional items on indirect costs/cost-sharing and on nondepartmental organized research units.
1993 Optional items on indirect costs/cost-sharing and on nondepartmental organized research units.
1994 Optional items on indirect costs/cost-sharing and on funds passed through to subrecipients.
1995 Optional items on indirect costs/cost-sharing, on funds passed through to subrecipients, and on cutoff value for reporting current fund research equipment expenditures.
1996 Added question on funds passed through to subrecipients to better address data discrepancy issue. Optional items on indirect costs/cost-sharing and on cutoff value for reporting current fund research equipment expenditures.
1997 Biomedical and bioengineering added as a seventh engineering subdiscipline. Included check-off item on whether R&D expenditures for medical schools were included in totals. Optional items on cutoff value for reporting current fund research equipment expenditures, R&D for non-S&E fields, and on federal R&D from specific federal agencies.
1998 Optional items on cutoff value for reporting current fund research equipment expenditures, R&D for non-S&E fields, and on federal R&D from specific federal agencies.
1999 Included check-off item on whether R&D expenditures for hospitals were included in totals. Optional items on R&D for non-S&E fields, on federal R&D from specific federal agencies, and on funds received as a subrecipient.
2000 Added question on funds received as a subrecipient to better address data discrepancy issue. Added optional items on R&D for non-S&E fields and on federal R&D from specific federal agencies.
2001 Included check-off item on whether R&D expenditures for clinical trials were included in totals. As a result of cognitive interviews with respondents and expert review of the survey form, several changes were made to the questionnaire layout.
2002 Made several changes to the web version to make it more user-friendly, clarify wording, and make it more compatible with the paper version.
2003 Removed the question on "burden hours" and the "optional" tags from Questions 2a (non-S&E R&D) and 2b (R&D from specific federal agencies).

- Highest degree granted
- Source of funds (federal, state or local, industry, institutional, or other)
- Type of academic institution (doctorate-granting or other)

These concepts and definitions are generally consistent in definition with the data collected from the federal government and industry, so they are capable of being combined with the results of the other surveys to get an aggregated view of the size, scope, and direction of R&D in the U.S. economy.

Field-of-Science Taxonomy

The interest in trends in college and university spending by fields of science arises largely in response to the need to trace the shifting emphasis in public funding for R&D by field. The most significant trend in funding by fields of science has been the reductions in funding for most fields of engineering and physical sciences, on one hand, and the accelerating growth in funding for biomedical research, on the other (National Research Council, 2001c). The recent report of the President's Council to Advance Science and Technology focused on the rather large changes in disciplinary areas supported by R&D funding over the past 25 years (President's Council of Advisors on Science and Technology, 2002). The report pointed out the rather large shifts in funding among science and engineering disciplines, such as in the physical and life sciences. As a base point: in FY 1970, support for the three major areas of research—physical and environmental sciences, life sciences, and engineering—was equally balanced. Today, the life sciences receive 48 percent of federal R&D funding compared with 11 percent for the physical sciences and 15 percent for engineering.

In order to assess these trends over time, it is necessary to get the data right; in order to get the data right, it is necessary to get the taxonomy of disciplinary classifications right and to apply that taxonomy consistently over time. This is not a simple matter. The taxonomy of fields of science has enlarged its purpose over the years, much as have all classification systems. In this case, a system that was designed to assist academia in sorting out doctoral programs in order to obtain reputational rankings of similar programs and to independently describe offerings has been pressed into service to support data collection in government and ancillary organizations. For the latter purpose the taxonomy is not so directly useful, so mechanisms for tracing activity by the classification system have not naturally evolved (Kuh, 2003). This evolution of the field-of-science taxonomy has also been the case with classifications of industry and occupation, which likewise have been pressed into uses for which they were not designed.

Like other classification systems, the classification by fields of science has its official sanction in a directive issued by the U.S. Office of Management and Budget (OMB). The most recent codification that delineates the "official" list of fields of science is contained in a directive entitled *Standard Classification of Fields of Science and Engineering*. This directive was last issued in 1978. In actuality, the standing list of fields has been around even longer, since the 1978 directive superseded without change a directive that had been published as OMB Circular No. A-46, dated May 3, 1974. It is uncertain how long that circular had been in gestation. In any event, the federal government is relying on a view of science and engineering that pertained some three decades ago.

In the ensuing years, much has changed in the science and engineering enterprise, as well as in the manner in which educational institutions organize their faculty and curriculum. Some of these changes were pointed out in a presentation at a workshop held by the Board on Science, Technology, and Engineering Policy by Charlotte Kuh of the National Research Council (NRC). In the dynamic biological sciences, for example, molecular biologists have been joined by biophysicists and structural biologists; genetics has been subdivided into molecular and general genetics; neurosciences now includes neurobiology; pharmacology now is pharmacology and toxicology; and so on. As science advances, new fields with new names spring up. Genomics morphs into proteomics, which may be regrouping into structural biology. These trends can only be identified in a system that permits field-specific variation and fineness of disaggregation, both of which tend to invalidate the cross-field and cross-data source comparisons to which the taxonomy has been addressed in recent years.

These organizational realities and the static nature of the taxonomy, it was reported in a previous National Research Council study, obscure some critically important information about the character of the science and engineering enterprise at colleges and universities:

- The diversity of some research fields, such as physics, which encompasses nuclear, particle, and solid state physics among the many subdisciplines.
- The growing importance of interdisciplinary and multidisciplinary research (discussed below).
- The changing emphasis within the fields, such as the shift to biologically based chemistry relative to physical chemistry within this field.
- The integration of some related fields, an example being the integration of electrical engineering, computer science, molecular biology, and biochemistry.
- The emergence of new fields and subfields, such as materials science, computational biology, and biophysics, and the near disappearance of others (National Research Council, 2001c).

There has been criticism over the years that the taxonomies are not standard from survey to survey. It is critically important to have comparable classifications among surveys so that the data can be compared and linked. Yet, as an example, the life sciences in the survey of college and university spending are disaggregated into agricultural, biological, medical, and other life sciences, while in the survey of federal support to universities, colleges, and nonprofit institutions, the life sciences disaggregation includes a fifth field—environmental biology (National Research Council, 2000).

The panel recommends that it is now time for the U.S. Office of Management and Budget to initiate a review of the *Classification of Fields of Science and Engineering*, last published as Directive 16 in 1978. The panel suggests that OMB appoint the Science Resources Statistics office of the NSF to serve as the lead agency for an effort that must be conducted on a government-wide basis, since the field classifications impinge on the programs of many government agencies. The fields of science should be revised after this review in a process that is mindful of the need to maintain continuity of key data series to the extent possible (Recommendation 6.1).

Non-S&E R&D Data

Many researchers in the fields of education, law, humanities, business, music, the arts, library science, and physical education would be surprised to find that the U.S. government does not consider their work to be part of science and engineering, particularly when so many of these fields have adopted investigative techniques and standards of evidence that replicate those of the "official sciences." Yet, by definition, these fields are excluded from the science and engineering totals.

There appears to be a widespread demand for collection and publication of data for non-S&E categories. This demand emanates from several sources. Survey respondents at institutions with a heavy concentration of research in the social sciences, education, humanities, the arts, and business were concerned that the survey did not give a complete and accurate picture of their research effort. In some schools, the data furnished to NSF did not jibe with the data in financial statements, which included non-S&E categories. The schools indicated that this would not be a large new burden, since many already collected the data for internal purposes and actually had to delete the non-S&E disciplines for the survey. NSF had its own reasons for initiating collection of these data; most countries followed the recommendation in the *Frascati Manual* to include social sciences (including educational sciences) and the humanities in the higher education R&D expenditure surveys. By excluding these data, the U.S. data were increasingly noncomparable to data from other countries.

To its credit, NSF has collected separate data on nonscience R&D spending since FY 1997. These data were collected on an optional basis, estimated, and used for internal purposes.

In recognition of the widespread demand for these data, NSF sought approval from OMB to collect these data on the surveys from FY 2003 through FY 2005, and with that approval, added these data to the regular questioning beginning with the collection of data for FY 2003. NSF plans to publish these estimates on a continuing basis.

Interdisciplinary Research

A major trend in the conduct of research and development in recent years has been an increase in the interdisciplinary nature of the enterprise. Collaborations between the fields of science, particularly between engineering and the other sciences, have characterized many innovations in R&D.

The current National Nanotechnology Initiative is an example. This multiagency project is pushing the frontier in nanoscale science and engineering that may have implications for medical diagnosis and treatment, efficient manufacturing processes, energy conversion and storage, and electronic devices. The applications in nanotechnology draw on advances in multiple disciplines, such as chemistry, physics, biology, and materials. The Office of Science and Technology Policy, which is coordinating this effort, points out that the initiative is blurring the distinctions of traditional scientific domains and creating a new culture of interdisciplinary science and engineering (National Science and Technology Council, 2003).

Thus, this is an example of a major initiative in which interdisciplinary research is being actively promoted by the federal government. Yet the NSF survey of R&D expenditures at colleges and universities not only fails to facilitate the collection of information to measure interdisciplinary research, but also, in fact, actively discourages respondents from reporting interdisciplinary research. The survey directs reporters to categorize interdisciplinary projects individually according to the nature of the research performance and to prorate expenditures to report the proportion of each discipline involved when multiple fields of S&E are encountered. It is even not possible for reporters to be creative and report interdisciplinary research as a subset of "other" fields, since respondents are directed not to use this category to report interdisciplinary or multidisciplinary research unless they cannot identify a primary field.[1]

The justification for and the basis of collecting information on the character and extent of interdisciplinary research was laid out by a National Research Council committee some years ago (National Research Council, 1987). In the realm of science education, the report defined *interdisciplinary* in terms of types of collaborations:

[1]"Other sciences not elsewhere classified" may be used when the multidisciplinary and interdisciplinary aspects make the classification under one primary field impossible.

1. Collaboration between different specialties of a discipline—for example, among cognitive, developmental, and social psychologists to understand the role of early childhood experiences in the comprehension of scientific concepts.
2. Collaboration among disciplines—for example, among chemists, educational psychologists, and cognitive scientists to structure chemical knowledge for effective instruction.
3. Collaboration among basic research, applied research, and development and application—for example, to improve the effectiveness of computer technology in the teaching of basic arithmetic operations.
4. Collaboration among practitioners, policy makers, and researchers—for example, to develop and adopt curricula and teaching materials that communicate scientific knowledge in a way that is both scientifically rigorous and educationally manageable.

The first two of these types of collaboration define interdisciplinary research of interest to the panel. **The panel recommends that NSF engage in a program of outreach to the disciplines to begin to develop a standard concept of interdisciplinary and multidisciplinary research and, on an experimental basis, initiate a program to collect this information from a subset of academic and research institutions (Recommendation 6.2).** A forthcoming NRC report on facilitating interdisciplinary research will also provide guidance on this subject (National Research Council, 2005).

Research Parks

University-related research parks have been identified as an important mechanism for the transfer of academic research findings, as a source of knowledge spillovers, and as a catalyst for national and regional economic growth.

In a November 2002 workshop convened to discuss policy needs for indicators related to the formation, activities, and economic consequences of research parks, the participants recommended that NSF learn more about U.S. research parks, their history and evolved relationship with university research, the roles of parks in the innovation system, the definition of park success, and the conditions for its achievement (Link, 2003). In his 2003 study for NSF, Albert Link inventoried and traced the development of 80 currently active university-related research parks across the country. These parks are considered to have contributed significantly to the advancement of science and engineering and the generation of innovation, and there is a strong movement to invest in these expensive assets by many governments and institutions. Yet the place of the research parks in the U.S. innovation system is not clearly understood.

Collaboration with Industry

The growing propensity of universities to enter into collaborative R&D arrangements with business and government laboratories has been a major trend in the R&D environment over the past two decades. There is evidence that universities are increasing funding links, technology transfer, and collaborations with industry and government. It has been postulated that these arrangements are expanding, in part, because of the decreasing role of the federal government as a funding source for R&D, encouraging universities to rely increasingly on nonfederal funding sources (Jankowski, 1999).

As discussed in Chapter 2, these collaborations have also been aided and abetted by government policies that encourage joint research activities, technology codevelopment, contract research, and technology exchange through licensing and cross-licensing arrangements (Vonortas, 2003). Some of the arrangements are underscored by equity arrangements and could be identified in equity transaction disclosures (such as reports to the Securities and Exchange Commission), but most are based on nonequity agreements.

It is critical that these arrangements be identified and understood in order to come to a complete understanding of the nature and extent of R&D activity in the economy. However, due to their nature, many slip under the radar screen of the dedicated databases that are designed to track the formal collaborations that have the blessing of law. The registry of research joint ventures under the National Cooperative Research and Production Act identifies only 9 percent of the 3,000 U.S.-based collaborations over the period 1985-2000 involving nonprofit institutions, including universities.

Some idea of the bounds of collaborations can be divined by reference to the total amounts that universities obtain from industry. But given the difficulties in obtaining a comprehensive and current view of increasingly important collaborations in their rich variety of forms, should the collection of R&D data from academia be extended to collect detail on types and amounts of collaborations?[2] Can the data of interest be obtained in this instrument that are of good quality without creating undue burden on the respondents? **The panel recommends that NSF consider the addition of periodic collection of information on industry-government-university collaborations as a supplemental inquiry to the survey of college and university R&D spending. A decision on the viability of this collection should be**

[2]The academic survey form now implicitly includes collaborative research ventures with other institutions in the expenditure totals, instructing respondents to "report only what your institution actually expends and accounts for when participating in joint research ventures." There is no attempt to separately identify and quantify these ventures.

preceded by a program of research and testing of the collection of these data (Recommendation 6.3).

Timeliness

In order to be fully relevant, the data must be available in a timely manner, so those who use them can compare them with other sources, aggregate them with other data to form a picture of the whole enterprise, and appropriately react to the information provided. The timeliness of the release of the academic survey data is slipping, not improving, over time. In the view of the panel, this renders the data less useful than they should be.

Some of the delays have to do with processing delays and other internal issues. The reporting forms are sent by mail and electronic means in November each year and are to be returned in January. A number of respondents need more time by virtue of the reporting processes and the timing of the fiscal year; they are offered an extension. This collection phase goes on until midsummer, when respondents who had not yet reported are asked to provide data only for items 1 and 2 (current R&D expenditures in S&E, by source of funds and by field of S&E). Total figures are asked for as a last resort. Respondents who still refused were thanked for their time and informed that NSF personnel might contact them in the future.

By the closing date of the 2002 survey in October 2002—some 9 months after the initial deadline for submittal of data—completed questionnaires had been received from 581 of the 610 academic institutions. These responses are tallied, data are edited, imputations are made, and estimates are prepared. Then the data are reviewed and prepared for release. These latter steps consume about half a year, so that the data are scheduled for release in April of the following year. In the natural order of things, the data would not appear until 18 months after the period of reference. NSF rarely meets that extended deadline.

Survey Methodology Issues

Several issues of survey methodology affect survey quality. In the survey design, all institutions in coverage with $150,000 or more in S&E expenditures reported in the previous survey are included with certainty. The frame is updated each year by comparing the previous frame for this survey with the list of institutions in coverage for the NSF-NIH Survey of Graduate Students and Postdoctorates in Science and Engineering and the list of academic institutions that receive federal S&E R&D funding as reported by federal agencies in the federal funds survey. The list of FFRDCs is maintained by NSF, and the list of historically black colleges and universities is maintained by the U.S. Department of Education. Thus, there is no sampling, as such, but errors may arise with the multisource frame itself and the application of the criteria for inclusion. For example, the frame could easily fail to identify an academic institution with at least $150,000 in R&D if it was not identified in the previous collection or found in the comparison with the graduate student survey. The elimination from coverage of less than 4-year colleges is presumed to result in only a very small underestimate.

For the past two collection rounds, the survey has been collected in electronic format only. It has usually been disseminated in November and follow-up activities have taken place from January to July, so the period of intensively soliciting responses consumes over 6 months.

Little is published or analyzed about the characteristics of the respondents in the institutions, although it is known that there is some annual turnover. The experience and knowledge of the respondents are critically important in this survey, since the appropriate completion of this fairly complex questionnaire usually requires obtaining data from multiple sources within an institution, as well as full understanding of such concepts and definitions as field of science and engineering. To maintain a check on the quality of reporting, for several years NSF has engaged the services of an expert in the field to visit institutions to report on the methodology used in completing the questionnaires, as well as issues relating to concepts and definitions (National Science Foundation, 2001, 2002). Likewise, veteran respondents are assembled by NSF to participate in periodic workshops to identify items that are troublesome. Despite these selective efforts to better understand the reporting, the panel is concerned about the lack of profile information about the respondents and the limited training or retraining of these respondents as part of ongoing survey operations. On an ongoing basis, **NSF should continue to contact a sample of responding institutions to check their records in order to improve understanding of the best means of gathering the data, the sophistication of reporting sources within an institution, and the interpretation of questions and definitions (Recommendation 6.4).**

It was noted that, by the closing date of October 2002 for the 2001 survey, completed questionnaires had been received from 95 percent of the academic institutions, including 100 percent of the top 100 institutions and FFRDCs. As was the practice with the ORC Macro surveys, all missing data items, including those for nonrespondents, were imputed. No item nonresponse rates were reported, so it was not possible to assess the quality of the individual items in the report. The printed collection form appears to be quite busy and suffers from a lack of good graphic design. This has been less of a problem in the electronic format, in which questions are interspersed with directions, definitions, and reminders. Concerned about the lack of knowledge about the response patterns, **the panel recommends study of the cognitive aspects of collection instruments and reporting procedures (Recommendation 6.5).**

Imputation is a significant issue for this survey. In 2001, imputation was used to provide information for a small proportion of the survey population (4 percent) that did not provide information at all, as well as for item nonresponse. The imputation factors were generated by class of institution and derived from responding institutions for three key variables: total R&D expenditures, federally financed R&D expenditures, and total research equipment expenditures. These factors were used along with the previous year's data. This methodology has led to large misspecification, especially when imputation is needed for a number of years. The estimates for basic academic research have been especially troublesome, since the response rate for this item has been in the range of 83 percent. In FY 2001, NSF needed to correct the "federal basic research" and the "total research" estimates by substantial amounts because of revisions in a large university's basic research spending—a number that had been imputed for 15 consecutive years.

NSF has conducted promising research to improve the imputation procedures. A memorandum by Brandon Shackelford of NSF outlined a promising approach for utilizing a regression model for imputation of the basic research totals, which has been subjected to initial tests (ORC Macro, no date). **Although the panel welcomes this research into a model-based approach to imputation, we are concerned that the tests were not sufficient to judge the soundness of the regression approach. The research should be redone utilizing a more standard procedure of withholding a set of independent data in order to test the model (Recommendation 6.6).** The past practice of using imputed data for long periods as a basis for new imputes is particularly dangerous.

SURVEY OF SCIENCE AND ENGINEERING RESEARCH FACILITIES

The objective of the annual survey of Science and Engineering Research Facilities is to provide data on the status of R&D research facilities at research-performing colleges and universities, nonprofit biomedical research organizations, and independent research hospitals in the United States that receive funding from NIH. The survey results have many uses, although NSF has determined that the predominant users are planners in the federal government, state governments, universities, and private-sector firms. Importantly, it has congressional interest, having been mandated by Congress in 1985.

Respondent Burden

This survey is one of the most burdensome surveys in the government's portfolio of surveys. Sometimes it is useful to put oneself into the shoes of the survey respondent when considering the appropriateness of a survey undertaking. Imagine you are an academic administrator with responsibility for completing federal government inquiries, and on your desk arrives a survey from NSF asking for the following information:

- The amount of space used for S&E research and how much of the space was for laboratories or offices.
- Condition of the research facilities.
- Costs of repairs and renovations in the last 2 years.
- New construction in the previous 2 years with a project worksheet to be completed for each individual project.
- Source of project funding for repairs and renovations and new construction.
- Planned repairs, renovations, and new construction for the next 2 years.
- Deferred repairs, renovations, and new construction.

Beginning in 2003, the survey goes onto a part 2 of the questionnaire, which contains new questions on computing and networking capacity. Questions asked are:

- The physical infrastructure used for network communications.
- Plans for future upgrading and uses of information technology.
- Capacity for high-speed computations.
- Infrastructure for wireless communication.

These exhaustive questions are garnered from a broad respondent base. All academic institutions granting master's or doctoral degrees in S&E or other institutions that reported R&D expenditures of $1 million or more, as well as all HBCUs reporting any R&D expenditures, are included. The six service academies are not included. Nonprofit biomedical research organizations and independent hospitals that received at least $1 million in extramural research funding from NIH in the previous fiscal year are also covered in the population of interest. (The frame for the survey is defined by the academic R&D expenditures for the academic institutions and an NIH list of nonprofit research institutions and independent hospitals that received funding from NIH.)

Although the amount of information sought in this survey has expanded and contracted over the years, the direction has been to add more rather than less burden over time (see Box 6-4).

To its credit, the NSF does a great deal to mitigate the burden and surprise factor accompanying this data collection. Data collection proceeds in three distinct steps. First, the president or equivalent of each institution is sent a letter signed by the director of NSF or the director of the National

BOX 6-4
Changes and Modifications to the Research Facilities Survey

1992 Most questions remained the same with the following additional questions: expenditures for central research infrastructure; expenditures for small repair/renovation projects; planned expenditures for animal research facilities; total space across all academic disciplines.

Two new topics were added to the survey: existence of an institutional plan and the number of years covered by the plan; costs, by S&E discipline, for needed new construction and repair/renovation.

1996 An optional question was added that asked for any additional comments from the institutions.

1998 Two new questions were added: list of nonfixed equipment costing at least $1 million; amount of indirect costs recovered from federal grants and/or contracts that was included in "institutional funds" if institutional funds were a source of funding for any new construction or repair/renovation.

In response to an OMB directive, a Large Facilities Follow-up Survey was added as a follow-up to the main facilities survey. Based on the responses to the main survey, institutions were sent a screening question and then a brief follow-up survey if they qualified.

1999 A new set of questions on animal facilities research space was substituted for the previous set of questions on this topic. The screening question for participation in the Large Facilities Follow-up Survey was incorporated into the main facilities survey.

2001 All but the first two questions of the survey were eliminated, as well as the Large Facilities Follow-up Survey. Generally, the survey asked for the amount of S&E instructional and research space by field of science and the adequacy of this space.

2002 OMB approval obtained to permanently discontinue the Large Facilities Follow-up Survey.

2003 A totally redesigned facilities survey questionnaire was fielded, including new and revised questions on "bricks and mortar" and on cyber infrastructure.

Center for Research Resources at NIH. This letter explains the survey and asks the president to name an appropriate person as the institutional coordinator (IC). A letter is also sent to the IC who had participated in the previous survey cycle, to notify the IC of the upcoming survey and the letter to the president.

The second phase is a mail-out of packages to the ICs, which contain a cover letter, an acknowledgment postcard to be mailed back, an overview of the survey, a paper copy of the questionnaire with a prepaid envelope, a list of frequently asked questions, and instructions for accessing the survey web site, with a user ID and a password. The NSF contractor, ORC Macro, maintains contact with the ICs about every other week.

The third phase consists of following up with the ICs. This means reminding them of due dates, answering questions, explaining how to use the web system, and so forth. About 75 percent of the ICs responded via the web.

NSF has also made several significant changes to sharpen the questionnaire. A cognitive study resulted in a rewording of the instructions and questions for greater clarity and some reformatting of the survey form. For example, some recent modifications include:

- The description of what was included in research space was expanded to eliminate incompleteness in the description.
- Redundant information from the instruction and information pages was removed.
- A choice of "not applicable (NA)" was added to the item asking for the amount of net assignable square footage used for instruction and research space, by S&E field (1a) and amount of the total research space for all S&E fields that was leased (1b). If a respondent answered "zero" to 1a, it was necessary to make "NA" a possible response for 1b.
- The new survey questionnaire also allows respondents to use data from their previous survey cycle data if they are still accurate for the current year. If so, processing staff entered these data for them on the paper questionnaire, and a "preload" button for each item accomplishes the same end for those who report on the web. These measures simplify reporting, but they could introduce errors to the extent that respondents take an easy way out of digging out fresh answers each year (National Research Council, 2002a).

A simple, but useful measure of the burden of a data collection is often reflected in the willingness of respondents to respond in a timely manner. Response rates for this survey have hovered around 90 percent for the academic institutions and 88 percent for the biomedical institutions. The top 100 (in terms of R&D expenditures) among the academic institutions had a 96 percent response rate; the private colleges had a lower but still acceptable 86 percent response rate.

Several reasons for these differences in nonresponse are postulated:

- Public institutions are accustomed to having their data public. Private schools do not have that tradition.
- Biomedical institutions are less likely than universities to have already collected these data in some form for a purpose other that responding to the survey. Also, these institutions are generally smaller than academic institutions. They may have received only one small grant from NIH. Thus, it is more of a burden to respond. Hospitals may not perceive the research grant to be as directly related to their main mission.

While it is tempting to suggest far-reaching changes to this admittedly burdensome survey, the panel notes that this survey has undergone extensive renovation in the recent past and has acceptable response rates from most survey respondent audiences. Since so many of the innovations in the questionnaire and in process automation are so recent, **the panel recommends that the experience in the fielding of the revised questionnaire in 2003 be carefully evaluated by outside cognitive survey design experts, and that the results of those cognitive evaluations serve as the foundation for subsequent improvements to this mandated survey (Recommendation 6.7).**

Survey Methodology Issues

The Survey of Scientific and Engineering Research Facilities is a biannual survey of academic institutions that includes a group of respondents not included in other NSF expenditure surveys—independent biomedical research facilities with NIH funding.

Survey Design

The eligible population has shifted somewhat over time. For example, the threshold for inclusion was raised from $150,000 to $1 million in 2003, and institutions granting master's degrees were no longer automatically included in 2001. The accuracy of the identification of the universe depends on the accuracy of the Survey of Research Development Expenditures at Universities and Colleges for the academic sector. Previously discussed errors in reporting and imputation may affect the quality of the list. There is a possibility of double-counting in the case of overlapping coverage between the NSF and the NIH lists.

Data Collection

The key to success of this survey is the institutional contact at each surveyed institution. For the most part, the input data are not maintained centrally within the institutions, so each institutional contact must determine the most effective data collection approach, work with multiple internal sources of information, and review the data before submission. The institutional contacts are identified in an intensive campaign preceding survey mail-out. They are not necessarily located in one type of office, and many of them change from year to year.

The questionnaire is evolving, with major changes introduced in the past two survey cycles. Many of the changes, the direct result of a cognitive review of the questions, were introduced to provide greater clarity and to remove redundancy. Our review indicates that this evolution process needs to continue.

The new design introduced with the web-based collection has increased the amount of data sought, introducing such questions as the identification of new construction in the previous 2 years, with a project worksheet to be completed for each individual project. Some of the concepts are new and possibly vague. For example, the new questions on computer technology and cyber infrastructure introduce new collection challenges, given the wide variety of institutional practices for computer and software procurement and inventory. **The panel recommends that NSF continue to conduct a response analysis survey to determine the base quality of these new and difficult items on computer technology and cyber infrastructure, study nonresponse patterns, and make a commitment to a sustained program of research and development on these conceptual matters (Recommendation 6.8).**

Even with these burdensome data inquiries, the overall response rate in 2001 was about 90 percent for academic institutions and 88 percent for the biomedical institutions. This is a testament to the institutional contact program and the determination of the data collectors. The differences in response rates between public and private institutions is of concern, with smaller rates for the private institutions perhaps the result of traditions and maybe the cause of larger error in their estimates. Some of these issues may be resolved in the current collection of data for 2003, when NSF will publicize data by institution, with only a few sensitive data items suppressed. The data should become more useful as benchmarks, and this procedure should also drive up institutional participation.

Processing

The survey employs web-based procedures, which require that all data items be completed prior to allowing submission of the form. This may force the respondent to enter doubtful data or to impute answers that are not obtainable from organizational records.

Imputation procedures vary for paper-based responses by whether or not the institution previously reported. If the unit previously reported, prior responses were used in the imputation procedure; if not, other methods were employed. **The exact procedure used by NSF for imputation is not well documented, but it appears that imputation is used for unit nonresponse—a practice that is highly unusual in surveys. In most surveys, unit nonresponse is handled by weighting, as it was in this survey in 1999. At a minimum, NSF is urged to compare the results of imputation and weighting procedures (Recommendation 6.9).**

7

Analysis of the R&D Data Funding Discrepancies

One characteristic of the portfolio of R&D expenditure surveys is that each survey was designed to meet a specific goal or set of goals. A previous National Research Council study concluded that each survey was developed to address a narrow topic rather than serve as an integrated part of a comprehensive system of R&D expenditure data (National Research Council, 2000). The nonintegrated nature of the portfolio came about as the survey managers scrambled to meet priority needs and initiated collections as resources became available, not necessarily as part of an overall master plan.

Despite much attention on the part of NSF to a concordance of concepts and definitions across the data collections, the independent development of the pieces of the R&D expenditure portfolio has led to gaps and anomalies that are of some trouble to data users. Perhaps no gap is more troublesome over time than the apparent discrepancy in funding and performance estimates among the surveys.

The topside measures of the apparent discrepancy illustrate this problem. When NSF collects data from the federal government on the amount that Congress authorizes to be spent on R&D (budget authority) that goes to academia and industry, it reports that the federal government has $86.8 billion to spend on R&D in fiscal year 2001. But when NSF tallies reports from academia and industry on funding received from federal sources during that same fiscal year, it only comes up with $72.6 billion. This discrepancy of $14.1 billion between budget authority and receipts is one depiction of the size of the discrepancy. Often, a comparison is made between obligations (which are tracked in detail in the federal funds survey) and performer receipts. The discrepancy between obligations and performer receipts in fiscal year (FY) 2001 amounted to a hefty $7.3 billion—smaller than the discrepancy between authority and receipts, but still worthy of attention. In an ideal world, the reports of spending and receipts would match, perhaps with some adjustment for lags in the transfer of funds.

There have been several attempts to disentangle this apparent discrepancy over the past several years (Congressional Research Service, 2000). This is considered an important issue by many. The discrepancy is annoying as an accounting anomaly to analysts who seek to trace the funding streams and answer the question "Where did the money go?" (Koizumi, 2003). It is also of concern to members of Congress and others who have oversight and legislative responsibilities, because the discrepancy casts a shadow over the accuracy of the data that are used to gauge the overall health and vitality of the nation's R&D enterprise, as well as answer the question "Are spending and receipt estimates accurately depicting how the money is flowing, and are the funds flowing in amounts and directions prescribed by legislation?" (U.S. General Accounting Office, 2001).

The discrepancy is of special concern to the panel and others who have studied it because its existence may signal a systemic quality problem in one or more of the NSF datasets. One litmus test of the quality of any statistical series is to compare the series with comparable data collected by other means. It is common statistical practice for federal statistical agencies to use administrative data sources to benchmark the totals derived from surveys. Measurement of undercoverage and overcoverage error (the failure to include or exclude some population units in the frame used for sample selection) often rely on comparing survey estimates to independent sources (U.S. Office of Management and Budget, 2001).

NATURE OF THE DISCREPANCY

As shown in Table 7-1, the discrepancy as measured by both obligations and outlays (see Box 2-1) has appeared to change course three times over the past three decades.

- Prior to 1980, the discrepancy was quite small and generally indicated that federal obligations exceeded performer spending.
- In the 1980s, the discrepancy became noticeable and changed, and for nearly a decade, spending appeared to significantly exceed obligations.
- Around 1990, the trend shifted again, and spending once again has lagged obligations and outlays, only now by fairly substantial amounts.

The timing of these swings in the data series is in close alignment with changes in the design and operation of the NSF data collections, so it is

TABLE 7-1 Comparisons of Federal- and Performer-Reported Expenditures for Federal R&D ($millions)

	Budget Authority (FY)	Total Obligations (FY)	Total Outlays (FY)	Total Performer Reported (CY)	Expenditures Less Budget Authority	Expenditures Less Obligations	Expenditures Less Outlays
1970	14,911	15,339	15,734	14,993	82	–346	–741
1975	19,039	19,039	19,551	18,615	–424	–424	–936
1980	29,739	29,830	29,154	30,002	263	172	848
1985	49,887	48,360	44,171	52,662	2,775	4,302	8,491
1990	63,781	63,559	62,135	61,596	–2,185	–1,963	–539
1991	65,898	61,295	61,130	60,799	–5,099	–496	–331
1992	68,398	65,593	62,935	60,815	–7,583	–4,778	–2,120
1993	69,884	67,314	65,241	60,581	–9,303	–6,733	–4,660
1994	68,331	67,235	66,151	60,787	–7,544	–6,448	–5,364
1995	68,791	68,187	66,662	62;965	–5,826	–5,222	–3,697
1996	69,049	67,653	66,142	63,341	–5,708	–4,312	–2,801
1997	71,653	69,827	68,898	64,548	–7,105	–5,279	–4,350
1998	73,569	72,101	70,632	66,340	–7,229	–5,761	–4,292
1999	77,637	75,341	70,585	67,015	–10,622	–8,326	–3,570
2000	78,664	72,863	69,807	66,371	–12,293	–6,492	–3,436
2001	86,756	79,933	75,336	72,637	–14,119	–7,296	–2,699

NOTE: Adjustments have been made to university reported R&D expenditures for 1998 and later years to eliminate double-counting of funds passed through one academic institution to another. Data for 1998 and later years are thus not directly comparable to data for 1997 and earlier years. For FY 1998, $479 million in pass-through funds were reported. For FY 2002, $865 million in pass-through funds were estimated.

useful to examine the possible relationship of these shifts with changes in survey design and implementation.

As the nature of the discrepancy suggests, there is probably no single point of origin of the difference between the data series. The discrepancy could arise from differing concepts, definitions, and interpretations in the data sources; from errors in the administrative reporting of federal funds; from measurement errors in the survey of R&D expenditures at universities and colleges; from sampling and measurement errors in the survey of industrial research and development; or from combinations of these sources of error. It has also been postulated that some of the discrepancy may be attributed to systemic reporting problems arising from the classified nature of some Department of Defense (DoD) R&D spending. In this section, we examine each of these potential sources of error.

Differing Concepts, Definitions, and Interpretations

In order for the data on federal obligations and performer expenditures to add up to the same total, they must be based on the same definitional criteria and have the same time horizon. However, federal obligations and performer expenditures are not definitionally the same and they do not necessarily pertain to the same point in time. In its report on the discrepancy, the General Accounting Office (GAO) pointed out that obligations are estimates of payments to be made by federal agencies without regard to when the payments may (or may not) ultimately be made. These obligations are made on the basis of the availability of appropriations and on the assumption that the performers will meet their expectations. Expenditures represent actual performer cost or expense data without regard to when the federal obligation was made. These expenses can occur years after the obligation (U.S. General Accounting Office, 2001).

Obligations are tied to orders placed, contracts awarded, services received, and similar transactions during a given period, regardless of when the future payment of money is actually required. The period for the reporting of obligations is standardized on the federal fiscal year: October 1 to September 30. Expenditures are reported when a reimbursement is made and in reference to a calendar year. These basic differences are exacerbated by federal contract payment practices. It was pointed out at the panel's workshop that no money comes to an academic contactor as a result of an obligation until the work is actually performed and billed, with a wait of 30-90 days for payment. Thus, at the performing institution, reported federal expenditures lag actual spending of money by several months. Since the average award is made toward the end of the fiscal year, this dynamic could lead to a natural delay between an obligation and an expenditure of 18 months to 2 years.

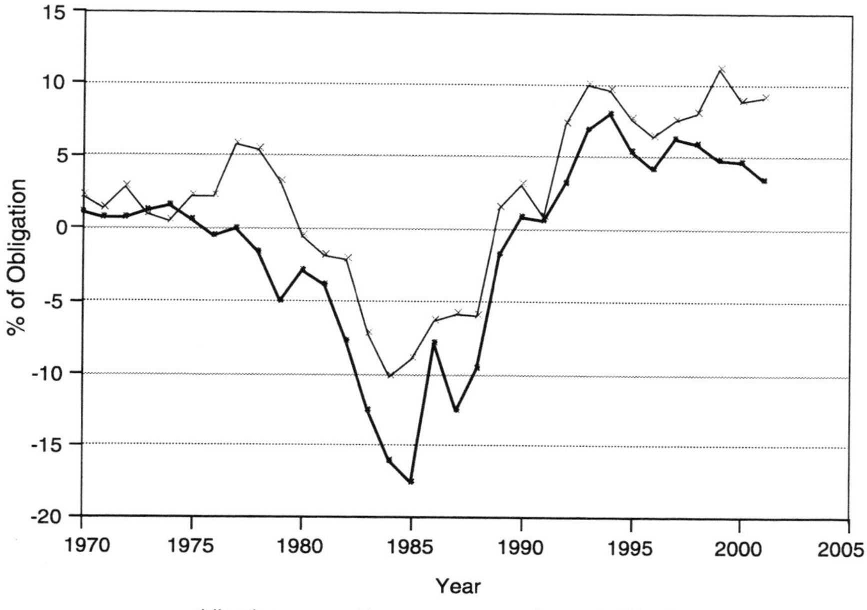

FIGURE 7-1 Comparison of annual federal and performer-related expenditures, 1970-2001.
NOTE: Data are shown as a percentage of annual obligations in order to adjust for inflation and changing level of federal support.

Considering these definitional differences, it is not surprising that there should be a discrepancy. The GAO report concluded that the discrepancy results primarily from annually comparing two separate and distinct types of financial data—federal obligations and performer expenditures—that are not comparable (U.S. General Accounting Office, 2001).

The extent of the contribution of this difference in definition and time horizon to explaining the alleged discrepancy can be seen when the comparison is made using outlays rather than obligations as the basis of comparison with expenditures. Outlays are generally considered to be closely akin to spending. They represent the amounts for checks issued and cash payments made during a given period, regardless of when the funds were appropriated or obligated. When these more comparable parameters are used as the basis for the reconciliation, a good part of the apparent discrepancy disappears. In 2001, when obligations exceeded expenditures by $7.3 billion, the difference between outlays and expenditures was just $2.7 billion (see Figure 7-1).

Although the size of this discrepancy varies by year, it could be said that the more valid comparison of federal outlays and performer expenditures reduces the discrepancy by one-half to two-thirds. On this basis, **the panel recommends that future comparisons of federal funding and performer expenditures be based on outlays versus expenditures, not obligations versus expenditures (Recommendation 7.1).** However, this more proximate alignment of definitions and time horizons still does not explain away the gap. Something else is apparently affecting the comparison.

Sampling and Measurement Errors in the Industrial R&D Survey

The most significant discrepancy resides between data reported by the federal government as obligated to R&D in industry and that reported by industry as expenditures in the industrial research and development survey. In this section, we examine the size and direction of this gap, summarize previous investigations into the source of this difference, and introduce discussion of a potential cause of the growing discrepancy over the 1990s.

As previously discussed, most of the overall discrepancy between federal obligations and reported expenditures has been between federally reported obligations to industry and the amounts reported as being received and expended by industry. The survey of federal funds reported $27.0 billion in obligations to industry in 2001, while the survey of industrial R&D reported that the federal government was the source of just $16.9 billion in industry expenditures. This gap of $10.1 billion, or nearly one-third, could be more than explained by the discrepancy between what DoD reports as R&D funding (obligations) and what industry reports it receives from DoD, a gap that amounted to $11.4 billion that year. Other agency-related gaps were smaller or offsetting.

A Congressional Research Service (CRS) study discussed several potential reasons for the phase shift in the gap in industry R&D numbers in the 1990s: acquisitions, mergers, and other structural changes in defense and aerospace industries in the decade; differing interpretations of R&D in the federal government and industry; and partial reporting by industry (Congressional Research Service, 1999, 2000).

When companies merge and acquire other companies, the quality and consistency of data reporting may suffer. In the 1990s there was an aggressive round of mergers, acquisitions, and other structural changes in the defense and aerospace industry that may have resulted in decreased awareness of the nature and scope of R&D in the new company. In some cases, smaller firms that were R&D intensive and contributed significantly to the estimates simply went away. Their R&D activities may or may not have been identified in the new reports of R&D data to the government provided by the acquiring company. This led to the conclusion that consolidation of the defense and aerospace industry contributed to the R&D data reporting problem (Quantum Research Corporation, 1999). Further on this issue, according to the CRS, DoD officials postulate that corporate mergers have affected continuity of reporting because job losses that accompany the consolidations have often pared individuals in front offices who had competently completed these surveys in the past (Congressional Research Service, 1999).

Although the definitions of R&D are reasonably clear and have been consistent over time, we have observed that there can be different interpretations of what constitutes R&D between a government agency and the companies that receive the funds at the individual project level. The Quantum Research Corporation (QRC) study compared a set of examples of project awards classified as research, development, test, and evaluation (RDT&E) by DoD with the categorizations provided by the recipient companies. The results of this small test were striking. Two-thirds of the companies did not consider most of the activities selected by DoD to meet their definition of traditional, reportable R&D work. A Census Bureau study of defense subcontractors came up with the same conclusion. Many of the subcontractors reported work as general R&D without ascribing it to an agency, often because the agency detail was not mandated by law (U.S. Census Bureau, 1997). This observation may well account for the fact that DoD expenditures are underreported by industry, while industry reports more "other" sources of funding than does the federal government. However, even if all of the gap in "other" were ascribed to DoD, it would account for only a minority of the DoD discrepancy.

Recalling that one of the so-called phase shifts in the gap between federal obligations for industry R&D and the amount reported by industry performers took place between 1991 and 1992, the panel examined results of the large-scale revision of the industry R&D survey that was introduced in 1991 to determine if that might have had an effect on the gap. Prior to the 1992 survey, the sample of firms surveyed was selected at irregular intervals, roughly associated with the economic census updates.[1] In the intervening years, only a panel of the largest firms known to perform R&D was surveyed. In latter years of the 1980s, for example, the sample was refreshed in 1987, when 14,000 firms were surveyed. From 1988 to 1991, just 1,700 of these firms were resurveyed, and the R&D data for the rest were estimated. This survey design biased the survey results in favor of large manufacturing firms, when it was increasingly apparent that there was substantial growth in R&D performance in the

---

[1]Prior to 1967, samples were selected every 5 years. Samples were selected then for 1967, 1971, 1976, 1981, and 1987.

nonmanufacturing sector and among smaller firms. Consequently, in 1991, the survey was redesigned to draw new samples with broader coverage annually and increase the sample size to approximately 23,000 firms each year. The impact of this change in the sample design is seen in the revisions to the data for the immediately prior year, 1990. The new sample design accounted for $13.7 billion or about 93 percent of the total $14.7 billion revision to total R&D, with most of that revision ($11.4 billion) attributed to better coverage of nonmanufacturing industries. In keeping with standard practice when dealing with large revisions, the new levels were wedged back in time, in this case to 1987, the year in which the last full sample was surveyed.

Although new estimates were prepared for 1991, it is obvious that a good part of the jump of nearly $5 billion in the size of the discrepancy between agency R&D spending and industry performer reports was attributable to changes in survey design. Some of the increase that year may be due to a one-time dip in DoD R&D spending, probably associated with diversion of R&D funds to support the Persian Gulf war effort, but the fact that the discrepancy continued to grow over the next couple of years, when DoD spending returned to prewar levels, indicates that a long-term shift in the data series indeed occurred.

The impact of the 1991-1992 survey redesign is readily apparent, but the question remains, why? In this matter, we can only speculate that the larger representation of small firms and nonmanufacturing industries changed reporting patterns. Outside the manufacturing sector, where R&D is conducted in a very formalized setting usually involving laboratories and R&D management structures, there may be less knowledge of the source of origination of R&D funds and more frequent pass-through and collaborative arrangements, which might well confuse respondents when asked to report on funding from federal sources. This may lead to underreporting among these firms. When the results of the underreported federal R&D receipts in the nonmanufacturing sector are weighted by the employment in that sector, the resulting estimate will fall short of an estimate derived largely from a survey that previously covered mainly the manufacturing sector. This downward bias in reporting of federal receipts appears to date back to the 1991 redesign of the industrial R&D survey. As the weight of the manufacturing sector in the survey has continued to decline with the growth in nonmanufacturing, this effect may well be the source of a large part of the growing reported discrepancy.

Errors in the Administrative Reporting of Federal Funds

In Chapter 5, we discuss several sources of potential error in the administrative reporting of federal funds in the annual Survey of Federal Funds for Research and Development. In this section, we highlight several of those sources of potential error, culled largely from previous studies by CRS and GAO as a means of pinpointing possible reasons for the discrepancy.

There is considerable evidence that timely and accurate completion of this survey is not a high priority for the federal agencies that annually provide the data. Top agency officials do not seem to understand the purpose and importance of collecting obligation data, in addition to the authority and outlay data provided regularly on a priority basis to the U.S. Office of Management and Budget (OMB) and to Congress on a "real-time" basis. Consequently, agency reporters may not receive sufficient resources to fully provide complete and timely data to NSF.

The lack of timeliness of data submission is indicative of the problem. NSF broadcasts its call for data in February of each year, closely on the heels of the release of the president's budget, requesting that agencies submit the data by May 15. Only about half of agencies meet that deadline, most blaming the requirements for extraordinary data compilation to levels not called for in managing the R&D programs nor in reporting to OMB. Labor-intensive and time-consuming processes without direct payoff to the agency are likely to be error-ridden, no matter how conscientious the agency reporters who have responsibility for furnishing the data.

Further complicating the ability to provide data to NSF have been dramatic changes in procurement practices in the federal government over the past decade. The shift away from specific task-oriented R&D contracts with the ultimate performer to larger contract awards which tend to be multiyear, multiagency, "blanket" indefinite order-type contracts and a trend toward "bunching" smaller contracts into large vehicles may have caused a difference in how R&D activities are categorized by respondents. These shifts in the composition of RDT&E procurements during the past 10 years were labeled the "most likely cause of divergence between the NSF reports" in a study by the Quantum Research Corporation (1999). However, a subsequent CRS study reported that DoD officials did not agree with the assessment that this was a primary cause, since many of the procurement reforms proceeded the growth in the gap (Congressional Research Service, 1999).

There is some speculation that agencies may not be consistently reporting R&D, despite the fact that NSF provides survey respondents with definitions of each data item collected on the surveys and stands ready to provide extensive hands-on assistance in ensuring that the right data are collected. Many of the agency reporters are budget analysts without detailed knowledge of the R&D enterprise and procurement activities in their agency. When confronted with judgment calls (e.g., should training, program evaluation, and construction be included?) they may not be as qualified to make the judgment as scientists and engineers and procurement officials (Congressional Research Service, 1999, 2000). Furthermore, even when agencies agree on definitions, they may be at variance with performer practices. Federal agencies often consider program management as an R&D activity, but performers may not (U.S. General Accounting Office, 2001).

Another source of potential error identified by participants in the 1998 NSF workshop on federal R&D was the challenge of determining the ultimate performer of R&D, particularly in larger R&D agencies with many of the aforementioned multitask, multiyear procurement vehicles. Even when the agency is able to identify the ultimate performer, the multitask contracts may mingle procurement and R&D activities in ways that make it difficult for the ultimate performers to appropriately break down the contract into its procurement and R&D pieces to prepare their NSF report. It has been observed that agencies tend to overreport intramural funding for federal laboratories, while underreporting R&D funding spent at an industrial or academic institution (Congressional Research Service, 1999, 2000). In addition, the visibility of some R&D funding appears to disappear when sent to another agency for dispersal. There may be inadequate mechanisms for interagency exchange of information that allows tracing of R&D data to the ultimate performer.

The changing nature of federal programs as they move through time is a further potential complication. An emerging program often moves through a progression from basic research to applied research to development to implementation, and the reporting system may not catch up with that progression. An example used quite often in the literature is the ambitious National Aeronautic and Space Administration (NASA) space station program, which saw significantly increased reported R&D obligations during the 1990s and sharply reduced reported R&D in 2000 when the program matured from an R&D-intensive activity to other categories of reporting. Industry-reported NASA R&D expenditures did not correspondingly reflect this upsurge and reduction in R&D, so the gap widened over the 1990s to the extent that, in 1998, industry reported a level of spending only one-third that reported by NASA (see Table 7-2).

Measurement Errors in the Survey of Academic R&D Expenditures

Compared with the size of the gap in the reporting of federal R&D outlays to industry and industry-reported R&D expenditures, the gap between reports of the federal government and academic institutions is minuscule (Table 7-3). In 2001, the difference between federal R&D obligations and reported university R&D expenditures amounted to only $397,000, or 2.0 percent of the total.

When adjusted to eliminate double-counting of funds passed through from one academic institution to another, the difference was a bit larger, totaling $1.1 million, or 5.6 percent. This adjustment for double-counting

TABLE 7-2 Selected Differences in Agency Reported R&D Funding to Industry and Industry-Reported R&D Expenditures ($millions)

	Total Federal R&D Obligations to Industry (FY)	DoD R&D Obligations to Industry (FY)	Industry-Reported DoD R&D Expenditures (CY)	DoE R&D Obligations to Industry (FY)	Industry-Reported DoE R&D Expenditures (CY)	NASA R&D Obligations to Industry (FY)	Industry-Reported NASA R&D Expenditures (CY)	Other R&D Obligations to Industry (FY)	Industry-Reported Other R&D Expenditures (CY)	Industry Total R&D Less Agency Total R&D
1981	16,282	10,931	10,540	2,486	1,976	2,096	2,306	769	1,560	100
1983	18,521	14,670	14,571	2,263	1,949	902	2,150	686	2,010	2,159
1985	23,496	19,069	20,948	2,348	2,901	1,312	NA	767	NA	NA
1987	28,628	24,258	22,252	2,128	2,968	1,479	2,805	763	2,732	2,129
1993	31,670	23,865	15,044	2,540	2,961	4,112	435	1,153	4,369	-8,861
1995	31,438	22,409	13,867	2,586	3,547	4,687	2,054	1,756	3,974	-7,996
1997	32,548	24,190	12,603	2,083	2,505	4,770	2,022	1,505	6,798	-8,620
1999	33,230	24,700	11,650	2,052	2,209	4,566	1,469	1,912	7,207	-10,695
2000	28,836	22,345	11,142	1,678	1,455	2,645	1,328	2,168	5,193	-9,718
2001	27,006	21,386	9,959	1,041	1,253	2,659	1,071	1,920	4,616	-10,107

NOTE: All years but 2001 include industry Federally Funded Research and Development Center (FFDRC) R&D. Beginning in 2001 these FFRDCs were no longer included in the industry R&D survey frame. CY = Calendar Year, FY = Fiscal Year, DoE = Department of Energy, NASA = National Aeronautics and Space Administration.

SOURCES: Data were derived from National Science Foundation, Division of Science Resources Statistics (NSF/SRS), Survey of Research and Development in Industry; and NSF/SRS, Survey of Federal Funds for Research and Development.

TABLE 7-3 Comparison of University Federal R&D Expenditures with Agency-Reported R&D Obligations ($millions)

	University R&D Expenditures			Expenditures Less Obligations		Percent of Federal Obligations	
	Reported	Net of Pass-through R&D	Federal Obligations	Reported	Net of Pass-through R&D	Reported	Net of Pass-through R&D
1988	8,193		7,828	365		4.7	
1989	8,991		8,672	319		3.7	
1990	9,638		9,138	500		5.5	
1991	10,234		10,169	65		0.6	
1992	11,093		10,271	822		8.0	
1993	11,957		11,208	749		6.7	
1994	12,646		11,797	849		7.2	
1995	13,328		11,928	1,400		11.7	
1996	13,836		11,978	1,858		15.5	
1997	14,309		12,559	1,750		13.9	
1998	15,145	14,724	13,381	1,764	1,343	13.2	10.0
1999	16,071	15,569	14,959	1,112	610	7.4	4.1
2000	17,508	16,878	16,815	693	63	4.1	0.4
2001	19,191	18,484	19,588	–397	–1,104	–2.0	–5.6

NOTE: Adjustments have been made to university-reported R&D expenditures for 1998 and later years to eliminate double-counting of funds passed through from one academic institution to another. For the purpose of comparison, both the reported federal R&D and the federal R&D net of federal R&D reported as passed through to another academic institution are shown.

SOURCES: Data were derived from National Science Foundation, Division of Science Resources Statistics (NSF/SRS), Survey of Academic Research and Development Expenditures; and NSF/SRS, Survey of Federal Funds for Research and Development.

has been made to university R&D since 1998 and was enabled by the addition of questions to the data collection form that break out how much of the R&D expenditures were passed through the institution to subrecipients, and how much were received by the institution as a subrecipient. These questions were added when NSF determined, as the result of an investigation, that at least $350 million of the $1.9 million difference in 1996 was due to double-counting (National Research Council, 2000). The effect of this correction was dramatic, although not necessarily consistent in direction (see Figure 7-1).

RECONCILING THE ESTIMATES

It is tempting to suggest that differences in definitions, time horizons, data collection methodology, and the effect of sampling and measurement errors simply invalidate a direct comparison of federal spending and performer expenditures. The estimates of federal funding for R&D in academia and industry, academic R&D expenditures, and industry R&D expenditures are independently derived from very different sources and cannot be expected to add up. However, it is essential to continue to peel away these potential sources of the discrepancy as a means of aiding in the interpretation of the NSF data, as well as a means of assisting in identifying sources of error in the estimates.

Thus, **the panel's recommendation is that a reconciliation of the estimates of federal outlays for R&D and performer expenditures be conducted by NSF on an annual basis (Recommendation 7.2).** This reconciliation should be published and widely disseminated by NSF as an aid to data users and as a blueprint for modifying the structure and implementation of the data collections to improve their concordance over time. The recent decision on the part of NSF to attempt to reduce a source of noncomparatibility by collecting data on potential double-counting of university R&D subcontracting was in direct response to attention to the size of the discrepancy. This change in data collection indicates that shining the spotlight on these differences can lead to improvements in data collection that, in turn, will reduce the gap over time.

8

Survey Management and Administrative Issues

The programs of the National Science Foundation's (NSF's) Science Resources Statistics Division (SRS) transcend their relatively modest size and their rather specialized audience in importance. In a way, these programs may be seen as a harbinger of the model of a statistical agency that many are propounding these days. That model envisions a federal statistical agency with a small staff of federal employees consisting primarily of subject-matter experts, supported by a limited number of specialists in survey management, survey design, the cognitive sciences, information technologies, and data analysis. In this model, survey operations are outsourced to other organizations that have a comparative advantage due to their size and concentration. The agency funds and otherwise manages the program and communicates its intentions by promulgating guidelines and standards and specifying performance in contracts.

SRS does not conduct its own statistical collections. Although other statistical agencies in the federal statistical system also outsource their data collection programs, the organization may be seen as a testbed for management of statistical programs in a nonstandard, resource-constrained environment. With a skeletal staff, SRS is expected to meet accepted standards of any federal statistical agency. Those standards pertain to a list of practices that include continual development of more useful data, openness about the data provided, wide dissemination of data, cooperation with data users, fair treatment of data providers, commitment to quality and professional practice, an active research program, professional advancement of staff, and coordination and cooperation with other statistical agencies (National Research Council, 2001b). The challenges it faces in meeting several of these standards with regard to the R&D statistics programs raise a cautionary flag to those who may propose that statistical agencies can thrive without conducting their own data collections.

INDEPENDENCE

To meet the needs of decision makers and other users of statistical data, a statistical agency must be able to provide objective, reliable information that may be used to evaluate programs and policies. To do so, the agency must have a widely acknowledged reputation for independence, which engenders trust in the accuracy and objectivity of the data. Unlike in many countries, where the independence of the statistical agency is organizationally ensured by virtue of the placement of that agency at a cabinet or subcabinet level, statistical agencies in the United States are part of mission-oriented departments and agencies. Within these departments, the independence of the statistical agency derives from its distinction from the parts of the department that carry out enforcement or policy-making activities.

The SRS division is fairly deeply embedded inside NSF, given that the agency has a fairly flat structure for a government agency (see Figure 8-1). Since 1991, the reporting arrangements have SRS as one of four divisions in

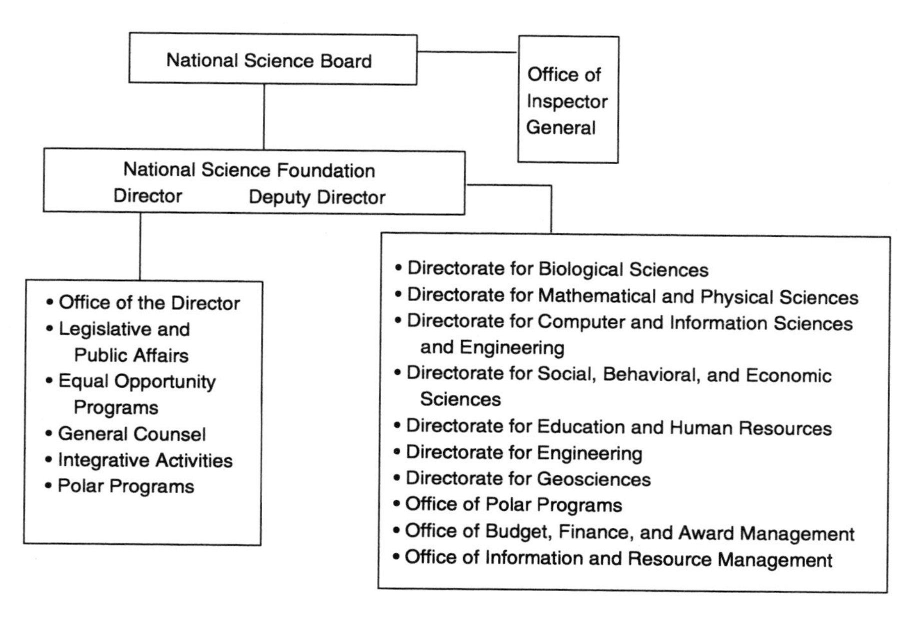

FIGURE 8-1 National Science Board organization.

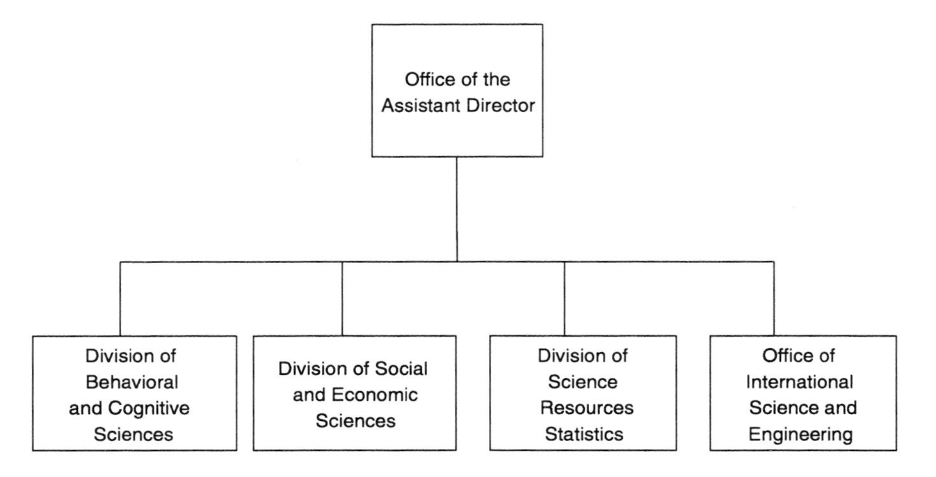

FIGURE 8-2 Directorate for Social, Behavioral, and Economic Sciences.

the Directorate for Social, Behavioral, and Economic Sciences (see Figure 8-2). This directorate is headed by an assistant director of NSF and has responsibility for two mission-oriented divisions as well as SRS and the Office of International Science and Engineering.

The mission of SRS extends beyond the mission responsibility of social, behavioral, and economic sciences. The division informs the operations of the Office of the Director and the Directorate for Social, Behavioral, and Economic Sciences. SRS also serves as an appendage of the National Science Board (NSB), certainly with regard to the preparation of *Science and Engineering Indicators*, and ultimately with regard to many other aspects of the mission of the NSB. However, the NSB does not have direct reporting units, instead conducting its activities through the National Science Foundation organization (see Figure 8-1).

There is no indication that this structure has had a negative influence on the independence of SRS. Within this structure, SRS has authority, if not autonomy, over decisions on the scope, content, and frequency of the data compiled, as well as for the selection and promotion of professional, technical, and operational staff.

The current reporting structure generates challenges in the area of resourcing. The SRS division budget is not separately identified as it winds its way through the directorate to NSF to the Office of Management and Budget (OMB) and to Congress, hence it competes, often unfavorably, with the important mission areas of the directorate and NSF. The separate reporting of the budget of the SRS Division in the OMB compilation of statistical programs of the U.S. government is in a sense an artifact, in that the data on SRS do not derive from a line in the administration's budget (U.S. Office of Management and Budget, 2003c). This lack of definition in the budget process can work to the detriment of an agency with programs that serve the needs of organizations and activities outside the home base of that agency, some of which derive directly from the interests of congressional committees.

The panel could find no compelling reason to suggest that SRS be relocated organizationally within NSF. However, we have the sense that an elevation of the visibility of the resource base for SRS would be a positive step and would serve to direct attention to the needs of the programs for sustainment and improvement (Conclusion 8.1). The recent elevation of the budget of the Office of International Science and Engineering, which has its own advisory committee and is only tangentially connected to the Directorate for Social, Behavioral, and Economic Sciences (SBE), can serve as a model. The budget for this organization has been identified as a separate line in the NSF budget.

ORGANIZATION AND STAFFING

SRS is organized functionally, with four major branches, each bearing responsibility for discrete program areas. Two of the branches have responsibility for data collection, analysis, and publication programs—the Research and Development Statistics branch has responsibility for programs addressed in this report, and the Science and Engineering (S&E) Education and Human Resources Statistics branch has responsibility for the human resources programs (see Figure 8-3). The two other branches are directed toward internal and external support—Information and Technology Services and Science and Engineering Indicators.

The statistical practices of SRS have been strengthened in recent years, as is shown by the issuance of NSF-SRS Quality Guidelines in 2002. However, since SRS contracts out all of its survey work, it is left with a very thin staff. There is little or no opportunity for growing a bench of expertise in the necessary methodological specialties, including mathematical statistics, cognitive sciences, and survey design, such as exists in the larger statistical agencies. This expertise, which can respond to new demands on the data, new methods of collecting and issuing data, and new methodological innovations, must be imported. Similarly, the day-to-day pressures of survey management and the recurring demands of intensive projects, such as the biennial preparation of *Science and Engineering Indicators*, limits the ability of the division to deploy staff to the task of data analysis. Data analysis requires access to respondents and respondent record-keeping practices, microdata in pure forms prior to aggregation into estimates, and the tools for production of the estimates. Since these tasks are contracted out and the

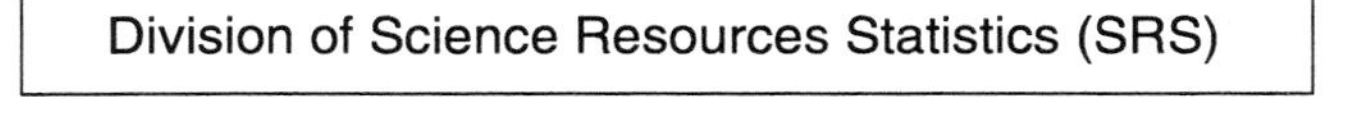

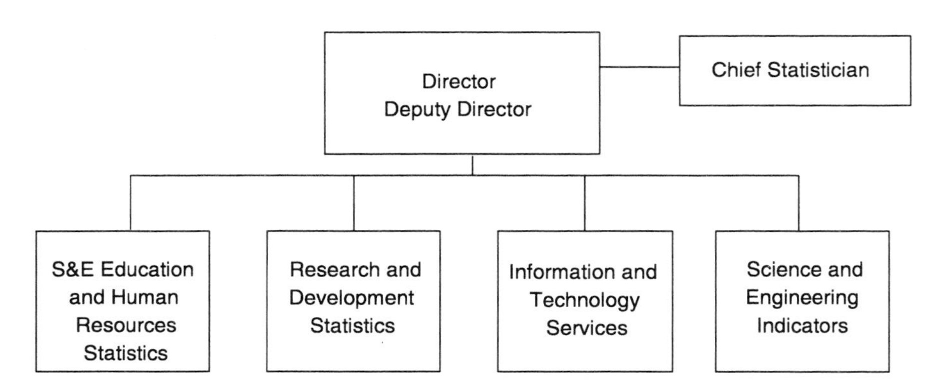

FIGURE 8-3 Organization of the Division of Science Resources Statistics.

microdata from the Census Bureau are treated in a confidential manner, the division has not been able to develop a needed analytic capability for understanding the changing character of research and development. As a result, it has not been able to develop internal competence in how to revise and review the surveys in the changing environment.

In this regard, the panel reiterates the concern of the previous National Research Council (NRC) report that concluded that SRS lacks the staff to plan and test the methods and instruments necessary to respond to the major needs of data users (National Research Council, 2000). That report recommended three courses of action to alleviate the staffing and skill shortages:

- augment staff expertise through professional development activities.
- develop a long-range staffing plan to bring on new hires with skills in statistical methods and data analysis.
- develop a more interactive relationship with external researchers to provide data and analytical support for the division's range of professional capabilities.

Although some progress has been made in each of these areas in the past 3 years, particularly with the recruitment of additional expertise in statistical and cognitive methodology, as well as by obtaining the services of nationally recognized outside experts in these fields for consultation and teaching engagements, there has not been sufficient time and resources to significantly enhance the data analysis skills.

DATA QUALITY

The panel concludes that SRS has the rudiments of a data quality program and further that SRS has adopted organizational and administrative changes to foster an environment for data quality.

We are hopeful that these recent initiatives, buttressed by additional resources and supplemented by further initiatives such as those outlined in this report, will lay a basis for further improvements in the future.

One important aspect of this quality improvement program is the development and promulgation of the publication *Statistical Guidelines for Surveys and Publications* in 2002 to set a framework for ensuring data quality. The guidelines are intended to (1) increase the reliability and validity of data, (2) promote common understanding of desired methodology and processes, (3) avoid duplication and promote the efficient transfer of ideas across program areas, and (4) remove ambiguities and inconsistencies.

The standards set goals, and the SRS surveys attempt to meet these goals. Yet there is inconsistent application of the guidelines. The guidelines contain a checklist to evaluate different types of nonsampling errors in the surveys. The use of the checklist in the different surveys is sporadic. Some surveys describe some aspects of survey methodology and omit other aspects. For example, none of the surveys adequately describes all of the aspects of unit nonresponse from the checklist: description of characteristics of units not responding, examination of changes in response rates over time, intensive follow-up of a sample of nonrespondents to ascertain nonresponse bias, and assessment of the method of handling nonresponse imputation or weighting procedures. Nonresponse is difficult to assess, but it is not impossible.

Because SRS has an interagency agreement with the Census Bureau to carry out the industrial survey of R&D, there are quality components written into the contract. The two tasks that are most pertinent to data quality are: (1) to prepare and maintain survey methodology and technical documentation and provide this information in a comprehensive methodology report and (2) to recommend quality improvements and methodological enhancements resulting from information gained in the previous survey cycle.

The panel notes that the interagency agreement with the Census Bureau covering survey operations for calendar year 2004 requires that the Census Bureau prepare and maintain complete and detailed survey methodology and technical documentation, and further that it recommend quality improvements and methodological enhancements resulting from information gained during previous survey sampling and processing.

The panel is concerned that, although the Census Bureau does provide a detailed methodological report every year, the information is quite uneven. Processing procedures, including complete editing procedures, are not fully described. The role of the data analyst in changing or adding data is not discussed. Imputation procedures are mentioned, but item imputation rates are not provided. Other important areas, such as the sample design, are fully documented and set the standard for making necessary improvements in overall documentation.

Census Bureau and SRS staff work together to identify areas of the industrial survey that need improvement. One recent, ongoing collaboration, noted in Chapter 3, concerns the provision of more stable estimates of state R&D. However, there are other areas that need improvement, such as a better understanding of the respondents and their difficulties in responding. Several recommendations made over a decade ago by a team of cognitive researchers at the Census Bureau have not been addressed. With the urging of OMB to develop a web-based questionnaire, an opportunity exists to do cognitive testing and a codification of processing and editing rules.

The panel urges SRS to take the lead in the work on the industrial survey, using the tools of the interagency agreement, the oversight of a high-quality methodological staff, and the input of highly qualified outside experts. This lead role would be undertaken while working collaboratively with the Census Bureau (Recommendation 8.1).

In addition to strengthening the overall quality and relevance of the survey, a more active lead management role by NSF will certainly enhance the subject matter and statistical expertise of the NSF staff, as well as make them more mindful of the strengths and weaknesses of the data. This model, in which the sponsoring agency plays an active role in decision making and directing, while leaving day-to-day survey management to the experts at the Census Bureau, is employed in various degrees in the relationship of the Census Bureau with other sponsoring agencies, most notably, the Bureau of Labor Statistics, with positive results for the surveys.

ANALYSIS

Currently, SRS does almost no analysis except for the biennial input to *Science and Engineering Indicators*. The detailed reports are issued with many statistical tables, and the *InfoBrief* series, designed to quickly release the data, are issued with few. Most of the analysis in the abbreviated publications consists of table reading.

The panel observes that the survey data are almost unexplored because there is a lack of staff with the expertise and the time to look at the data intensively, explore issues raised by users, juxtapose data from different sources, and generally learn the strengths and weaknesses of the data in a substantive setting. There are no publications of high caliber in which staff data analysis can be readily published.

Lacking analysis, SRS loses an opportunity to learn more about the methodological weaknesses of the data. The methodology-analysis cycle is one of the strongest means of improving survey data. Until an agency uses its own data extensively, beyond descriptive statistics, it does not really understand the gaps and needs in its collection system. The panel recommends that SRS engage in research on substantive issues that would lead to improvements in the design and operation of data collection and to a fuller understanding of the data limitations. The panel further recommends that, over time, SRS develop both the internal staff capacity in data analysis and a suitable vehicle or vehicles for professional publication of data analysis by both internal staff and outside experts.

MEASURES OF EFFECTIVENESS

The discussion of the strengths and weaknesses of SRS has so far focused on the uniqueness of the agency, the organizational structure, and the principles and practices that must be in place for SRS to fulfill generally accepted minimum requisites of a federal statistical agency. These factors are inputs to a quality product. Several measures exist to judge performance, or the output of the agency. Among these are timeliness, relevance, and balance. Each of these output measures is discussed in turn.

Timeliness

It is axiomatic that data are more useful in a current than a historic context, and they are more useful when issued close to the reference period for the survey than some time after it. One of the most frequent complaints of potential data users about the R&D data is that the data are released too late. As a result, users rely on data that are less comprehensive, of lower quality, and, frequently, quite old when the analysis is performed.

The issue of timeliness of the data has been raised in several critiques of the R&D data over the years. The previous NRC study in 2000 concluded that "SRS must substantially reduce the period of time between the reference date and data release date for each of its surveys to improve the relevance and usefulness of the data" (National Research Council, 2000:5).

If anything, the recent trend has been to lengthen rather than shorten the time from reference period to publication. For example, the data from the Survey of Industrial Research and Development for calendar year 2000, which had a December 31, 2000, reference period, were not available in their full publication detail until June 2003, some 30 months after the reference period. A partial release of six tables was made in February 2002, and the *InfoBrief*, with some narrative table reading, was available in April 2002.

This 30-month period contrasts unfavorably with the timeliness experience for the industrial R&D survey as reported by SRS for the previous 6 years, since the large-scale redesign of the survey between 1991 and 1992 (data displayed by reference calendar year in months from reference month to publication of full detail):

1993	1994	1995	1996	1997	2000
19	17	23	15	13	30

The data for the past few years for all of the major SRS surveys show that things have become worse, not better, since the previous NRC panel recommended improvement in timeliness. Table 8-1 traces the timeliness of the release process, as each R&D expenditure survey and the national patterns report moved from the reference period, to date of early release (month/year of posting on the web), to *InfoBrief* posting on the web, and to the posting/publication of the full set of detailed statistical tables for the past 4 years.

For every survey, there has been slippage and, in most cases, at least 2 years slippage. This lack of timeliness adds to the underutilization of the data.

Some of the delay is in the processing and provision of the data by the Census Bureau and the contractor for the web-based surveys. The Census Bureau has recently set a goal to release the industry survey by the end of the calendar year in which the survey questionnaire was distributed. For example, it is investigating processing changes so that 2004 data are released to NSF by the end of calendar year 2005. NSF would still be responsible for assembling the data release to the public.

The industry survey is the largest of the R&D surveys. If it can improve timeliness by earlier mail-out and increased automation of the disclosure review, perhaps the four smaller web-based surveys can also find ways to accelerate data release.

Relevance of Program Content

The relevance and utility of the data are expected to increase simply by their earlier provision. Still, there are major substantive gaps in the R&D

TABLE 8-1 Survey Release Schedules

	Reference Period	Early Release Tables	*InfoBrief*	Full Set of Tables
Survey of Industrial Research and Development	Dec-99	Jan-01	May-01	Apr-02
	Dec-00	Jan-02	Dec-02	Jun-03
	Dec-01	NA	Jun-03	Dec-03
	Dec-02	NA	Jun-04	NA
Survey of Federal Funds for Research and Development	Sep-99	Dec-00	Mar-01	Jul-01
	Sep-00	NA	Feb-02	Jun-02
	Sep-01	NA	Jun-03	Apr-04
	Sep-02	NA	May-04	NA
Survey of Federal Science and Engineering Support for Universities, Colleges, and Nonprofit Institutions	Sep-99	Jan-01	Mar-01	Apr-01
	Sep-00	Feb-02	Feb-02	May-02
	Sep-01	NA	Mar-03	May-03
	Sep-02	NA	May-04	NA
Survey of Research and Development Expenditures at Universities and Colleges	Oct-99	Dec-00	NA	Jul-01
	Oct-00	Dec-01	NA	Feb-02
	Oct-01	NA	Aug-03	Apr-03
	Oct-02	NA	May-04	Jul-04
Survey of Science and Engineering Research Facilities	Sep-99	Feb-01	NA	Jul-01
	Jun-01	Jan-02	NA	Mar-02
	Sep-03	NA	NA	NA
National Patterns of R&D	1999	NA	Oct-99	Nov-99
	2000	Nov-00	Nov-00	Mar-01
	2002	NA	Dec-02	Mar-03
	2003	NA	Mar-04	Aug-04

NOTE: If the milestones do not apply (e.g., there were no early release tables, there was no *InfoBrief* written, or the full set of tables were not published), NA is noted.

program that have been discussed throughout this report. At the present time, there are many unanswered questions. For example, what is the impact of academic research, both nationally and regionally? How do you measure innovation's contribution to economic growth? What is the entire globalization picture? There is some information on research in the United States funded by companies abroad and some information about research abroad funded by the United States, but there is no overall picture of the entire process. Similarly, there is no accounting for R&D done in interdisciplinary centers, and interdisciplinary work is a growing trend.

In addition to the gaps in the substance of the program, there are gaps in the methodological development of the surveys. Four of the five surveys discussed in this report are done primarily on the web. Yet very little, if any, methodological work has been done to find out how paper and web data collection can support each other. Except for the two federal surveys, the other surveys all use imputation. It is handled in very different ways, with no one comparing and suggesting alternative methods, and no one measuring the impact on the estimates and, when appropriate, their variances.

The industrial R&D survey is the most complex of the five surveys, as well as the most costly. In 2003, its cost was $1.2 million. For a survey of this importance, it is not getting the attention it needs either from SRS or the Census Bureau, which conducts it. For the Census Bureau, it is one of the smaller surveys it conducts. For that reason, it is used as a training ground for new employees, who then graduate to bigger surveys. For staff who do not move on, there is little incentive to suggest changes or improvements. The Census Bureau processes this survey much as it does the other manufacturing surveys it conducts. In the case of the R&D survey, this means a mail-out of the questionnaire to a company with no designated respondent, no means of compiling a response rate that reflects how many companies reported data, no information on item nonresponse, and an obscure editing process that is not documented. SRS does not have the staff it needs to manage the Census Bureau staff well and to insist that the survey be done to meet well-accepted statistical standards.

Balance

The panel has observed that SRS, in the press to produce its surveys, cannot devote sufficient attention to broader concerns of data quality and coverage. To maintain high standards of statistical quality and keep abreast of important trends in R&D, SRS should align its personnel so that internal experts can provide ongoing guidance on those matters without any direct responsibility for administering a particular survey.

Currently, many respondents do not seem to understand the definitions of the various items. SRS staff do not seem to understand what is being reported. Each survey should have at least one staff person who could discuss the items and definitions with respondents. This should be a proactive measure. In the press of production work, there is no offsetting balance for dealing with respondent concerns and therefore ensuring that the data collected are really reflecting current R&D status.

Because the R&D environment is changing so rapidly, there is an ongoing need to keep abreast of users' needs. Workshops should focus on users' needs and types of questions. There should be a staff person whose job it is to assimilate the information provided in workshops, from users, from a proposed advisory committee for the industrial survey, and from groups in other countries who work on similar surveys. At the present time, all effort is focused on production; there is very little thought given to improvements for the future.

In order to keep the R&D program relevant, both programmatically and methodologically, there should be a mechanism for proposing studies and analyzing research. There should be a mechanism for commissioning studies to deal with specific suggestions for problem areas. The constitution of a methods test panel for testing new methodology, new questions, and new technology could be of great value to SRS. However, someone needs to decide on the focus of such a panel—to analyze data, to work with respondents, to advocate change. Such a person or persons cannot be running a survey with its own pressing deadlines.

NSF should appoint two high-level experts with the following responsibilities: to help create a balance between ongoing effectiveness in issuing survey data and the need for tracking important trends in research and technology that might affect the respondent bases of surveys, to work with individual surveys on adapting to new requirements, and to serve as liaisons with outside experts, while addressing some of the long-term statistical issues at SRS. One should have expertise in statistics and one in economics. These people should not be saddled with administrative chores. Instead they should be devoted to innovation and exploratory work on the R&D surveys. Each of them should work on analyzing the data, one with a focus on meeting user needs, and one with a focus on providing the best methodology for the surveys.

TOOLS TO IMPROVE DATA QUALITY

Quality starts with an organizational commitment and is buttressed by professional standards of practice. These are two recognized elements that help define an effective statistical agency (National Research Council, 2001b). The commitment should involve several key elements: use of modern statistical theory and sound statistical practice, strong staff expertise, an understanding of the validity and accuracy of the data, and quality assurance programs that include documentation of concepts and definitions, as well as data collection methodologies and measures of uncertainty and possible sources of error. In view of the importance of the overall environment for quality, early in our work we reviewed these elements of organizational and management commitment to quality.

As a benchmark, the panel turned to the earlier NRC study *Measuring the Science and Engineering Enterprise: Priorities for the Division of Science Resources Studies* (National Research Council, 2000) and evaluated the progress that SRS has made to implement the recommendations for quality improvement in that report. This earlier report found that SRS had a good track record of improving the data quality and meeting statistical standards in the recent past, and it recommended additional steps to ensure that standards are met across SRS operations. With respect to the R&D expenditure surveys, the previous NRC report recommended three specific steps to improve data quality: (1) require all contractors to provide methods reports that address quality standards, (2) continue recent efforts to provide NSF staff with professional development opportunities to improve statistical skills, and (3) continue to develop and strengthen a program of methodological research, undertaking rigorous analysis of the data collected to assess the quality of the data relative to concepts they are supposed to measure.

Since the 2000 NRC report was published, SRS has implemented several specific improvements and has laid the groundwork for others. The change of name to the Division of Science Resources *Statistics* is one indication of the commitment of management to improving the statistical foundation of the R&D data. Other specific actions that have improved or will lead to improvements include the addition of mathematical statistics expertise to the SRS senior staff, the incorporation of specific contractual obligations for measuring and presenting measures of quality in contracts with the data collection organizations responsible for conducting the several survey programs, and development of a long-term plan for methodological research for each program.

Mathematical Statistics Expertise

The 2000 NRC report focused considerable attention on perceived staffing issues in SRS, calling for enhancing staff expertise through professional development activities, augmenting staff expertise in statistical methods and subject-matter skills, and developing a more interactive relationship with outside researchers. Over the past 2 years, SRS has added four full-time senior statisticians and has secured the services of experts in survey design and cognitive aspects of questionnaire design. These staff and consultant enhancements have served to focus attention on statistical methods for the R&D surveys.

Long-Term Plan for Methodological Research

The multiyear plan for methodological research on the survey of R&D in industry is perhaps the most advanced of these efforts to specify a roadmap for near-term and long-term statistical improvements. In a presentation at the panel's July 2003 workshop, Ray Wolfe of the SRS research and development statistics program outlined a research agenda and plan of action for the R&D industrial survey. The plan is specified in the SRS Quality Improvement and Research Project List. The list was developed with input from, among others, the OMB clearance process, the SRS staff, the NRC's Committee on National Statistics and the Board on Science, Technology, and Economic Policy, and SRS advisory committees.

The Quality Improvement and Research Project List that was current at the time of the panel's workshop included 30 items, 9 of which were presented in some detail to the panel. Three of the research items were designed to address OMB clearance conditions: record-keeping and administrative records practices, effects of mandatory versus voluntary reporting on the state items, and web-based survey operations. Other items on the list of nine key activities include research on data collection below the company level, cognitive research on survey forms and instructions and identification of appropriate respondents, survey sample design issues, survey processing (data editing, cleaning, and imputation), research on disclosure and confidentiality issues, and electronic documentation system for respondent contacts. Each of these research agenda items is discussed in this report.

In particular, the panel encourages SRS to develop plans to review and resolve several cross-cutting issues of statistical methodology. These issues are web-based collection, provision of prior-year data to survey respondents, the designation of respondents, and nonresponse adjustment and imputation.

Web-Based Collection

The four federal and college and university surveys rely mainly on the Internet to support data collection. (A few respondents are not able or willing to respond electronically, so a manual system is also maintained.)

Many of the traditional chores of survey operations are eliminated in web-based collection. The survey is delivered electronically, rather than in paper format through the mail. The respondent completes an electronic survey form. Follow-up is by e-mail or telephone rather than postal mail. Logs of progress of the agencies and institutions in reporting are generated by the system electronically. The web-based questionnaire enables embedded edits that question the entry of erroneous data, so that data checks, repairs, and explanations can be provided by the respondents and the data are tabulated by the web-based system. In this family of surveys, the target population is small, and the processing environment is simplified in that there is no sample, no nonresponse adjustment by weighting, and no other weighting.

The very newness of the web-based system and the employment of automated functions raise several red flags in terms of data quality. For one thing, the system can force respondents to provide data that they are unsure about, simply to complete the entry. In this sense, it forces imputation by the respondent, rather than by NSF. In more traditional survey approaches, the failure to complete a data item may signal a problem with questionnaire design, concepts, or definitions. These useful signals are masked by self-imputation. Another consequence of forcing an entry is that traditional measures of nonsampling error, such as item nonresponse and imputation rates, may be artificially low; statements in NSF publications that "item nonresponse is not viewed as a problem" because item nonresponse rates are under 1 percent may be quite misleading as a measure of overall quality.

Against this backdrop, OMB has challenged NSF to explore the possibility of web-based collection in the larger and more stylized Survey of Industrial Research and Development. The terms of clearance of the survey, when it was last approved by OMB in January 2002, require that "NSF will initiate a web-based version of this survey by the next submission of this clearance request. If NSF is unable to complete a web-based version upon resubmission, NSF must provide justification as to why web-based surveys can not be implemented in this instance, and document the steps being taken to allow full electronic submission of this survey" (U.S. Office of Management and Budget, 2002:2).

NSF and the Census Bureau have introduced several technological initiatives for data collection over the years. For example, firms using the short form (RD-1A) that report no R&D activity can fulfill the reporting requirement by the use of toll-free touch-tone data entry. Nearly 90 percent of such respondents in 1998 and 1999 used the touch-tone system. In addition, the Census Bureau utilizes a computerized self-administered survey system, which allows respondents to complete the survey form electronically and submit it on a disk. Approximately 20 percent of firms that were sent both a paper form and a diskette in 1999 chose to report using the diskette. Although these initiatives offer reduced respondent burden and facilitate data processing at the Census Bureau, they fall short of representing the data collection and processing improvements offered by the kind of web-based collection utilized in other NSF surveys. In essence, the current initiatives represent merely an automation of the old way of doing business.

Clearly, an intensive regime of development and testing of further automation of data collection, processing, and estimation is required. In keeping with the OMB recommendation that research should focus on web-based collection from businesses, a proof-of-concept study was conducted by the Census Bureau in the 1990s (U.S. Census Bureau, 1998). Work on web-based collection should consider the dynamic aspects of questionnaire design and processing discussed in a recent Committee on National Statistics workshop on survey automation (Tourangeau, 2003). Such a study needs to be developed and conducted in SRS. A comprehensive feasibility study for an effective web-based instrument requires both use of interactive contact with individual respondents, as recommended in this report, and a carefully designed field pilot study.

Provision of Prior-Year Data to Survey Respondents

Many of the NSF surveys provide data collected in prior-year responses on the current year collection instruments. The industry survey imprints prior-year data on the RD-1 form and incorporates the information on diskettes forwarded to the companies under the Census Bureau's computerized self-administered survey option. For the other surveys, prior-year data are available to respondents on request or automatically.

There are potential strengths and drawbacks to presenting historical data for a current reference period. A paper by Holmberg (2002) outlines motives that include, on the positive side, increasing the efficiency of the data collection, correcting previous errors, reducing response burden, and possibly reducing measurement errors and spurious response variability. Arrayed against these advantages are the possible drawbacks that preprinting can conserve errors rather than correct them, and that the practice might lead to underreporting of changes from one period to another. Holmberg's experimental study of preprinting values on a self-administered questionnaire for business establishments by Statistics Sweden found that questionnaires with preprinted historical values outperformed questionnaires without preprinted values on several measures of data quality. These data quality improvements included (not surprisingly) a significant decrease in response variation in overlapping reference periods, a reduction in spurious reports of change, and less coding error and editing work. He further pointed out that some of the potential drawbacks were not realized (for example, the entry of the same value for different months was more common without preprinting), and there was no indication that change was underestimated with preprinting.

The favorable results evidenced in Holmberg's research suggest that there is little danger in continuing this practice in the NSF surveys. However, there is the possibility that his results would not be replicated in U.S. surveys conducted with the spectrum of methods used in the NSF surveys. The panel notes that the Census Bureau has initiated testing of the impact of preprinting on its Survey of Residential Alterations and Repairs (personal communication, D.K. Willimack, U.S. Census Bureau, September 2, 2003).

The panel urges NSF to sponsor research into the effect of imprinting prior-period data on the industrial R&D survey in conjunction with testing the introduction of web-based data collection (Recommendation 8.2). The traditional means of testing the impact of differential data collection procedures using randomized split samples and respondent debriefings should be considered for this research.

Designation of Respondent

A major challenge in all of the NSF surveys is to ensure that the survey questionnaire will get to the appropriate and knowledgeable (and motivated) respondent. The role of the data provider is central to all of the surveys, and it is handled very differently from survey to survey.

The Survey of Science and Engineering Research Facilities identifies one person as an institutional coordinator. The survey recognizes that this person is critical to the success of this survey, which must be transferred among various parts of the institution. The institutional coordinator decides what others should provide, what sources of data to use, and who coordinates the pieces of the effort, then reviews the data. This plan appears to generally work well for this survey.

In sharp contrast, the Survey of Industrial Research and Development is mailed to a company, with no contact person designated. Often the form is sent to a central reporting unit at corporate headquarters, with respondent "generalists" fanning out the collection to various organizations and functions within the corporation. There is little attempt to develop a standing relationship with these key reporters or to educate them on the nuances of the data collection.

The role of the data provider is critically important to the success of a survey. We note that, in response to discussions at a meeting of panel members with NSF and data collection staff in April 2003, it has been proposed that the NSF Quality Improvement and Research Project List research agenda will document research contacts for the industry R&D survey.

The panel supports the initiative to identify individual respondents in companies as a first and necessary step toward developing an educational interaction with respondents so as to improve response rates and the quality of responses (Recommendation 8.3).

Nonresponse Adjustment and Imputation

The surveys have very different approaches to nonresponse and imputation. The federal funds and college and university surveys attempt to eliminate the problem of unit nonresponse completely, by seeking 100 percent compliance from the universe of reporters. Item nonresponse is actively discouraged, either by making it difficult not to enter an item in a web-based report, or by encouraging reporters to estimate information when actual data are not available, as is the practice on the federal funds report. Thus there is little unit or item nonresponse in this family of surveys, although, as mentioned earlier, some of the methods of achieving these response rates have a questionable overall impact on data quality, and their impact needs to be investigated in an evaluation study.

The Survey of Industrial Research and Development has fairly significant response challenges. Of the companies surveyed in 2000, about 15 percent did not respond. Moreover, the means of counting and then treating nonresponse in this survey raises statistical issues. The Census Bureau lists a company as responding when it has heard back from the company, even if the company reported that it was out of scope, out of business, or merged with another company. In other words, a response is a form returned, not necessarily information received. The Census Bureau then imputes values to data items representing these nonresponding units. The panel is concerned about this procedure, in that it does not follow acceptable practices for reporting nonresponse, and we are concerned about the impact of this practice on both the reported nonresponse rates and the estimates. The reported values for item nonresponse rates for key data elements in the survey are also quite problematic. Imputation rates are published, but they are a poor proxy for item nonresponse rates.

The panel recommends that NSF, in consultation with its contractors, revise the Statistical Guidelines for Surveys and Publications to set standards for treatment of unit nonresponse and to require the computation of response rates for each item, prior to sample weighting (Recommendation 8.4). When these guidelines have been clearly specified, the panel expects that the Census Bureau and other consultants would adopt these standards. With clear specification by the NSF and adoption by the contracting organizations, information on true unit and item nonresponse can be developed and assessed.

Survey Documentation

The methodological reports from the various surveys vary widely in their completeness. The SRS guidelines are quite clear on the policy for survey documentation in a methodological report expected at the completion of each survey. There is some evidence that recent methodological reports have improved in terms of depth and focus (see, for example, ORC Macro, 2002). The means of ensuring continued improvement in survey documentation is specified in the SRS guidelines—that is, survey contracts must include a requirement for a methodology report covering items addressed in this standard (National Science Foundation, no date).

Resources

The SRS budget is small for a statistical agency. Only about $3.7 million was allocated for carrying out the R&D expenditure surveys in

2003. None of the surveys has even as many as two full-time-equivalent (FTE) staff on the payroll of NSF. At this level of funding and staffing resources, it will be difficult, if not impossible, for NSF to implement a program of research and survey improvements that are recommended in this report.

In keeping with the thrust of the recommendations of this and previous NRC panels, the highest priority in the direction of available resources should be given to redesigning the industrial R&D survey. In particular:

- Update the questionnaire to sharpen the focus of the survey and fix the problems identified in this report. This should be conducted in conjunction with a program of research on web-based collection and tested by means of a panel on small methods development.
- Carry out a thorough investigation of sampling for the industrial survey, considering the use of other frames and multiple frames.
- Research record-keeping practices of companies to determine if line-of-business data could be collected on the industry survey.

Across-the-board improvements will be made only when NSF augments its own internal staff expertise in order to exercise greater control over the content and operations of the surveys—a process that has begun in the agency but on a limited basis. The panel thinks that in order to facilitate the recommendations in this chapter, SRS should take the following steps:

- Augment the staff with the services of high-level experts who can think ahead for the R&D surveys, filling gaps, improving methodology, and analyzing data. These could be additions to the staff but, probably would be drawn from outside (academic) experts in the subject-matter fields.
- Implement greater control over the conduct of the surveys by rigorous utilization of the contract and cooperative agreement vehicles.

On a longer-term basis, NSF can engage in a deliberate process to redesign, or at least revitalize, all of its surveys on a rotating schedule. Such a program will have many components, some of which are suggested in the discussions of the individual surveys. NSF should fund research on questionnaire design for many of the surveys and on how to make the web design truly useful.

Some of these initiatives can be accomplished within the resource base of NSF, simply requiring a shift in emphasis and direction. Others will require a new infusion of resources into the SRS R&D expenditure programs. The requirement for new funding arises in a budget situation that has seen slow and unsteady growth over time. The panel hopes that the fairly significant increase in budget authority in fiscal year (FY) 2003 can be preserved to enable SRS to make some improvements in some of the survey activities. The budget for FY 2004 was about the same as the FY 2003 budget, with a slight increase for methodological work for the industrial R&D survey. We note that the costs of the nonindustry surveys are proportionately much higher. Although the surveys that collect data on federal, academic, and nonprofit R&D are sent to many fewer respondents, the surveys are more complicated than the industry survey, and response rates are generally higher, pushing up costs (see Table 8-2 for detailed expenditure information).

Advisory Committees

An advisory committee with rotating membership gives an agency a method to bring in experts in different areas, as needed, in both substantive and survey methodological fields. An advisory committee can listen to reports, respond to questions about processes, raise questions, suggest new questions, argue about gaps in the data, and make recommendations. At present, SRS has an advisory committee that is a break-out group of the Social, Behavioral, and Economic Sciences Advisory Committee. In view of the lack of a dedicated advisory committee, NSF has sought to obtain advice and guidance on content and structural aspects of its surveys from many sources over the years, some through standing bodies and others on an ad hoc basis. The panel believes that for all surveys, outside advice and consultation is necessary to provide expertise as well as to legitimatize the surveys.

The Survey of Industrial Research and Development in particular would benefit from a dedicated, focused, formal, and ongoing external advisory panel knowledgeable about these data and the issues they help inform. In the late 1990s, external survey-relevant advice was obtained using alternative mechanisms. In 1998, NSF provided funding to the NRC's Science, Technology, and Economic Policy Board to conduct a workshop assessing the utility and policy relevance of government's data on industrial research (including specifically the industrial R&D survey) and industrial innovation. The purpose of the workshop was to generate suggestions for improving measurement, data collection, and analysis. Workshop participants included statisticians and economists concerned with industrial organization and innovation practices, industrial managers, association representatives, government officials representing diverse policy arenas and statistical agencies, and analysts from international organizations and other industrialized countries. As a follow-up to the workshop, the Census Advisory Committee of Professional Associations, specifically the American Economic Association component, convened an R&D miniconference to recommend changes

TABLE 8-2 Sample Size and Expenditures for NSF Surveys

Survey Year	Sample Size	Response Rate %	Mode	Survey Costs	Redesign & Related Costs	FTEs
All R&D surveys						
1993				$1,289,000	$433,000	7.50
1994				920,000	459,000	7.50
1995				1,698,000	—	7.50
1996				2,102,000	—	7.50
1997				2,580,000	—	7.50
1998				1,708,000	—	7.50
1999				2,216,000	—	7.50
2000				1,709,000	—	7.50
2001				2,259,000	200,000	7.50
2002				2,668,000	200,000	8.25
2003				3,687,000	600,000	8.50
Survey of Industrial Research and Development						
1993	24,064	81.8	Paper	390,000	433,000	1.50
1994	23,541	84.8	Paper	405,000	459,000	1.50
1995	23,809	85.2	Paper	575,000		1.50
1996	24,964	83.9	Paper	801,000		1.50
1997	23,417	84.7	Paper	783,000		1.50
1998	24,809	86.6	Paper	850,000		1.50
1999	24,431	83.2	Paper	855,000		1.50
2000	25,002	84.8	Paper	876,000		1.50
2001	24,198	83.0	Paper	894,000	200,000	1.50
2002	31,200		Paper/web	1,197,000	200,000	1.75
2003	31,200		Paper/web	1,200,000		1.75
Survey of Federal Funds for Research and Development						
1993	105	100.0	Disk	149,000		1.50
(sample size includes subagencies)						
1994	102	100.0	Disk	139,000		1.50
1995	98	100.0	Disk	170,000		1.50
1996	94	100.0	Disk	185,000		1.50
(costs exclude database task)						
1997	93	100.0	Disk	226,000		1.50
1998	92	100.0	Web	267,000		1.50
1999	90	100.0	Web	209,000		1.50
2000	93	100.0	Web	315,000		1.50
2001	74	100.0	Web	170,000		1.50
2002	74		Web	379,000		1.75
2003	74		Web	334,000		1.75
Survey of Federal Science and Engineering Support to Universities, Colleges, and Nonprofit Institutions						
1993	15	100.0	Disk	90,000		1.50
(sample size includes major agencies only)						
1994	15	100.0	Disk	147,000		1.50
1995	15	100.0	Disk	169,000		1.50
1996	18	100.0	Disk	148,000		1.50

Survey Year	Sample Size	Response Rate %	Mode	Survey Costs	Redesign & Related Costs	FTEs
(costs exclude state profiles task)						
1997	20	100.0	Disk	164,000		1.50
1998	19	100.0	Web	243,000		1.50
1999	19	100.0	Web	194,000		1.50
2000	18	100.0	Web	168,000		1.50
2001	18	100.0	Web	125,000		1.50
2002	18		Web	397,000		1.50
2003	18		Web	334,000		1.50
Survey of Research and Development Expenditures at Universities and Colleges						
1993	681	96.9	Paper/disk	375,000		1.50
1994	500	99.6	Paper/disk	229,000		1.50
1995	499	90.3	Paper/disk	229,000		1.50
1996	493	97.3	Paper/disk	453,000		1.50
1997	493	98.0	Paper/disk	347,000		1.50
1998	556	98.6	Web/paper	348,000		1.50
1999	597	98.5	Web/paper	310,000		1.50
2000	623	97.3	Web/paper	350,000		1.50
2001	609	97.3	Web	620,000		1.50
2002	610		Web	695,000		1.75
2003	610		Web	552,000		1.75
Survey of Science and Engineering Research Facilities						
1993	309	93.0	Paper	285,000		1.50
(excludes biomedical sample)						
1994						1.50
1995	307	97.0	Paper/disk	555,000		1.50
1996						1.50
1997	350	87.0	Paper/web	545,000		1.50
1998						1.50
1999	556	73.0	Paper/web	648,000		1.50
2000						1.50
(two-question survey)						
2001	580	90.0	Web/paper	450,000		1.50
2002						1.50
(complete redesign)						
2003	585		Web/paper	1,267,000	600,000	1.75
Survey of Research and Development Funding and Performance at Nonprofit Institutions						
1996				515,000		
1997	9,112	41.4	Paper/web	515,000		

NOTES: Research Facilities Survey is a biennial survey. FTE totals exclude the nonprofit survey.

and provide guidelines for improving the analytical relevance and utility of the industrial R&D survey statistics.

The Survey of Research and Development Expenditures at Universities and Colleges benefits from ongoing annual site visits, respondent workshops, and advice received from its special emphasis panel. This survey employs an external consultant, Jim Firnberg, to make multiple site visits to survey respondents. He is well-known and respected by institutional respondents, having once been a respondent to the NSF surveys and now very active in several national university organizations, including the Association for Institutional Research. His findings and recommendations are included in an annual *Institutional Response Practices* report, which has been the basis for identifying topics that have been more fully investigated through advisory panels and academic workshops. An advisory special emphasis panel was established in the late 1990s. In 2001, the discussion topics for this panel included (1) survey difficulties resulting from an overlap of sector boundaries (including specific discussion of hospitals and clinical trials and of consortia and pass-through funding), (2) better accounting for indirect costs, and (3) the status of optional items on non-S&E R&D performance and federal agency-specific reporting.

Annual workshops are held with respondents to discuss issues related to the academic R&D survey. Workshop topics have included technical data preparation guidance, cognitive response issues, and policy relevance of the survey questionnaire and its content. The survey instrument has been revised (questions added and modified, instructions clarified) as a result of these workshops. Participants tend to be senior university budget or research administrators whose offices are responsible for the survey response. In 2002, for example, the workshop confirmed from smaller universities that NSF should include the optional items on non-S&E R&D and federal agency sources of support as core questions on the survey.

The Survey of Scientific and Engineering Research Facilities has benefited from the advice of a special emphasis panel (or expert advisory groups) since its inception in the late 1980s. Participants tend to be senior university facilities or budget office administrators whose offices are responsible for survey response, and nonacademic institutional representatives who share similar facility concerns (such as the Howard Hughes Medical Institute). An experts meeting in Chicago in 2003 was convened to identify whether it would be possible to collect information on cyber infrastructure expenditures.

Both the Survey of Federal Funds for Research and Development and the Survey of Federal Science and Engineering Support to Universities, Colleges, and Nonprofit Institutions benefit from respondent issue workshops. A formal special emphasis panel has not been established for these federal surveys. In 1998, in lieu of a formal advisory panel, a Federal

Agency Workshop on Federal R&D was convened. Workshop participants included senior officials from budget and policy offices of most major federal R&D funding agencies and from major data users (including OMB, the Congressional Research Service, the National Research Council, the Consortium of Social Science Organizations, and the American Educational Research Association). Proceedings included discussions on the federal science and technology budget, R&D data by fields of science and engineering, how major R&D-performing agencies use the federal funds survey data, and agencies' accountability for reporting R&D data.

NSF has convened a variety of multiagency respondent issue workshops to review the content and survey reporting process for the combined federal funds and federal support surveys. In addition, NSF has sponsored several agency-specific workshops to address technical issues of direct concern and relevance to individual agencies, the latest in 1999.

For the Survey of Research and Development Funding and Performance by Nonprofit Institutions, a special emphasis panel was established at the very outset to guide the development of survey content and sampling coverage. In 1996, this panel met to provide guidance and make suggestions for the planned 1996 and 1997 nonprofit R&D data collection effort. The panel consisted of 11 senior officials from the nonprofit, academic, and government communities with knowledge about, and an analytical interest in, the nonprofit sector.

Opportunities for Efficiencies

Although the SRS staff seems to be spread very thinly, there may be opportunities for efficiencies. One was mentioned earlier: that is, investigation of the use of a RaDiUS-like collection of contract, grant, and cooperative agreement data, either as replacement or an adjunct to the federal funds survey.

The academic R&D survey already makes use of the frame developed for the human resource surveys of the SRS division to identify doctorate-granting institutions. It may be that the creation of a database of universities and colleges, with distinguishing characteristics, would be a worthwhile investment that would reduce the annual cost of identifying the frame for the academic R&D survey. Such a frame would undoubtedly be useful for other purposes. Perhaps the blending, whenever possible, of the R&D and human resources surveys would create efficiencies.

Generic Clearance

SRS has recently received OMB approval for a 3-year generic clearance to conduct a broad range of survey improvement projects (pretests, case studies, pilot surveys, customer surveys, focus groups, and methodological research) and to establish a quick response capability. SRS has suggested four studies to begin the process: three focusing on cognitive testing of survey items and one to develop the quick response capability. With this generic clearance, SRS may be able to conduct more research in a timely way to improve survey content and methodology.

Mandatory Reporting

Item nonresponse in the R&D surveys outside the two surveys of the federal government is rampant. It is especially serious in the industrial survey, where many large companies refuse, as a matter of company policy, to furnish data unless the survey is mandatory.

In most years, the industrial survey is a blend of mandatory and voluntary items. From 1958 to 1960, only the item on "cost of R&D performed" was mandatory, and that was because it was deemed a requirement for the economic censuses. Then, in 1961, the "federal governments funds" portion of the cost of R&D performed became mandatory. Two more data items—"net sales and receipts" and "total company employment"—were added as mandatory items in 1969. These four items were the only mandatory items until, in 2001, "cost of R&D performed within the company by state" was made mandatory. For 2002, OMB approved that all items be mandatory, because of the 2002 economic censuses, but in 2003, there was a return to the five mandatory items. The Census Bureau has requested that all items be mandatory in the year of economic censuses; it has also proposed that the number of items be scaled back in interim years.

On a test during the 1990 survey cycle of the industrial survey, SRS looked at the effect of reporting on a completely voluntary basis. The sample was split in two. One-half was asked to report as usual on the mix of mandatory and voluntary items; the other half was asked to report on a completely voluntary basis. The result was a decrease in the overall response rate. The test showed that the mandatory items experienced a sharp decrease in response when voluntary reporting was permitted. However, no information is available on what would happen to voluntary items if mandatory reporting were enforced. The primary problem with the industry survey is the voluntary items, which can have very large nonresponse.

Neither SRS nor the Census Bureau know the reasons why companies do not report the voluntary items, although there are many suggestions. **The panel recommends increased reliance on mandatory reporting between economic censuses to improve data quality, reduce collection inefficiencies, and provide greater equity among reporters. However, the panel also recommends additional research on the topic of voluntary versus mandatory reporting, to investigate whether mandatory reporting is the most effective strategy (Recommendation 8.5).**

References

Achs, Z., and D.B. Audretsch

2003 Innovation and technological change. In *Handbook of Entrepreneurship Research*, Z.J. Achs and D.B. Audretsch, eds. Boston: Kluwer Academic Publishers.

Acs, Z.J., H.L.F. de Groot, and P. Nijkamp

2002 Knowledge, innovation and regional development. In *The Emergence of the Knowledge Economy*, Z.J. Acs, H.L.F. de Groot, and P. Nijkamp, eds. New York: Springer.

Adams, J.D., and A.B. Jaffe

1996 Bounding the effects of R&D: An investigation using matched establishment-firm data. *RAND Journal of Economics* 27(Winter):700-721.

Altshuler, R.

1988 A dynamic analysis of the research and experimentation tax credit. *National Tax Journal* 41:453-466.

Andersson, C., H. Lindstrom, and L. Lyberg

1997 Quality declaration at Statistics Sweden. Statistical Policy Working Paper 26. Seminar on Statistical Methodology in the Public Service. Washington, DC: U.S. Office of Management and Budget.

Archibugi, D., and G. Sirilli

2001 The direct measurement of technological innovation in business. In *Innovation and Enterprise Creation: Statistics and Indicators*, European Commission (Eurostat), ed. Luxembourg: European Commission.

Archibugi, D., S. Cesaratto, and G. Sirilli

1991 Sources of innovative activities and industrial organization in Italy. *Research Policy* 20:299-313.

Arundel, A.

2003 *Patents in the Knowledge-Based Economy*. Maastricht, Netherlands: Merit.

Arundel, A., G. van de Paal, and L. Soete

1995 *PACE Report: Innovation Strategies of Europe's Largest Industrial Firms: Results of the PACE Survey for Information Sources, Public Research, Protection of Innovations, and Government Programmes, Final Report*. Prepared for the SPRINT Programme, European Commission. Maastricht, Netherlands: Merit, University of Limbourg.

Atkinson, R., and R. Court
1998 *The New Economy Index: Understanding America's Economic Transformation.* Washington, DC: Progressive Policy Institute.

Australia Bureau of Statistics
2002 Discussion Paper: Measuring a Knowledge-based Economy and Society—An Australian Framework. [Online.] Available: http://www.ausstats.abs.gov.au/Ausstats/free.nsf/Lookup/4F8E59034103E624CA256C230007DC05/$File/13750_aug%202002.pdf [accessed July 9, 2004].

Bean, A.S., M.J. Ruse, and R. Whitely
2000 Benchmarking your R&D: Results from IRI/CIMS Annual R&D Surveys for FY '98. *Research Technology Management* 43:(1):16-24.

Bostic, Jr., W.G.
2003 Census Bureau Perspective on Improving the RD-1 Survey. Unpublished paper presented at the Panel on Research and Development Statistics at the National Science Foundation's Workshop on R&D Statistics at NSF, July 24-25, Washington, DC.

Branscomb, L.M., and P.E. Auerswald
2002 *Between Invention and Innovation: An Analysis of Early-Stage Technology Development.* NIST GCR 02-841. Washington, DC: National Institute of Standards and Technology.

Brooks, H.
1968 *The Government of Science.* Cambridge, MA: MIT Press.

Bureau of Economic Analysis
1994 A satellite account for research and analysis. *Survey of Current Business*, November. [Online.] Available: http://bea.gov/bea/articles/NATIONAL/NIPAREL/1994/1194od.pdf [accessed April 1, 2004].

Bush, V.
1945 *Science—The Endless Frontier: A Report to the President on a Program for Postwar Scientific Research.* Washington, DC: Office of Scientific Research and Development.

Butani, S., K.W. Robertson, and K. Mueller
1998 Assigning permanent random numbers to the Bureau of Labor Statistics Longitudinal (Universe) Database. *Proceedings of the Section on Survey Research Methods* (American Statistical Association):451-456.

Cesaratto, S., S. Mangano, and G. Sirilli
1991 The innovative behaviour of Italian firms: A survey on technological innovation and R&D. *Scientometrics* 1:207-233.

Chesbrough, H.W.
2003 *Open Innovation.* Cambridge, MA: Harvard Business School Press.

CHI Research, Inc.
2002 *Small Serial Innovators: The Small Firm Contribution to Technical Change.* Haddon Heights, NJ: CHI Research, Inc.

Cochran, W.G.
1997 *Sampling Techniques.* Hoboken, NJ: John Wiley & Sons, Inc.

Cohen, W.M., and S. Klepper
1992 An anatomy of industry R&D intensity distributions. *American Economic Review* (September):775.

Cohen, W.M., R.C. Levin, and D.C. Mowrey
1987 Firm size and R&D intensity: A re-examination. *Journal of Industrial Economics* 35(4):543-565.

Cohen, W.M., R.R. Nelson, and J.P. Walsh

2000 *Protecting Their Intellectual Assets: Appropriability Conditions and Why U.S. Manufacturing Firms Patent (or Not)*. NBER Working Paper 7552. Cambridge, MA: National Bureau of Economic Research.

2002a Links and impacts: The influence of public research on industrial R&D. *Management Science* 48(January):1-23.

Cohen, W.M., A. Goto, A. Nagata, R.R. Nelson, and J.P. Walsh

2002b R&D spillovers, patents and the incentives to innovate in Japan and the United States. *Research Policy* 31(December):1349-1367.

Committee for Economic Development

1998 *America's Basic Research: Prosperity Through Discovery*. [Online.] Available: http://www.ced.org/docs/report/report_basic.pdf [accessed July 21, 2004].

Congressional Research Service

1999 *Explanation of Disparity in Industry Reported DOD R&D Expenditures*. Memorandum. Washington, DC: Congressional Research Service.

2000 *Challenges in Collecting and Reporting Federal Research and Development Data*, M. Davey and R. Rowberg, eds. Washington, DC: Congressional Research Service.

Corrado, C., C. Hulten, and D. Sichel

2004 Measuring capital and technology: An expanded framework. In *Measuring Capital in the New Economy*, C. Corrado, J. Haltiwanger, and D. Sichel. Chicago: University of Chicago Press (forthcoming).

Ducharme, L-M., and F.D. Gault

1992 Surveys of advanced manufacturing technology. *Science and Public Policy* (December):393-399.

Economic Development Administration

2001 *Strategic Planning in the Technology-Driven World: A Guidebook for Innovation-Led Development*. Washington, DC: U.S. Department of Commerce.

Energy Information Agency

2001 Technical Information on CBECS. [Online.] Available: http://www.eia.doe.gov/emeu/cbecs/technical_ information.html [accessed April 12, 2003].

Ernst, L., R. Valliant, and R. Casady

2000 Permanent and collocated random number sampling and the coverage of births and deaths. *Journal of Official Statistics* 16:211-228.

European Commission

2003a Innovation Policy: Updating the Union's Approach in the Context of the Lisbon Strategy. [Online.] Available: http://europa.eu.int/smartapi/cgi/sga_doc?smartapi!celexplus!prod!DocNumber&lg=en&type_doc=COMfinal&an_doc=2003&nu_doc=112 [accessed July 9, 2004].

2003b *Raising EU R&D Intensity*, Report to the European Commission by an Independent Expert Group. Luxembourg: Office for Official Publications of the European Communities.

European Council

2000 Presidency Conclusions, Lisbon European Council. [Online.] Available: http://ue.eu.int/ueDocs/cms_Data/docs/pressData/en/ec/00100-r1.en0.htm [accessed July 9, 2004].

Eustace, C.

2003 The PRISM Project: Report of Research Findings and Policy Recommendations. Report Series No. 2. European Commission Information Technologies Program. [Online.] Available: http://www.euintangibles.net/research_results/FinalReport.pdf [accessed April 8, 2004].

Federal Trade Commission
1985 Testing the Theory of Competition in R&D of Large U.S. Companies. Economic policy paper by J.T. Scott and G.A. Pascoe, Jr. Washington, DC: Federal Trade Commission.
Fossum, D., L.S. Painter, V. Williams, A. Yezril, E. Newton, and D. Trinkle
2000 *Discovery and Innovation: Federal Research and Development Activities in the Fifty States, District of Columbia, and Puerto Rico*. Santa Monica, CA: RAND.
Fossum, D., L.S. Painter, E. Eiseman, E. Ettedgui, and D.M. Adamson
2004 Vital Assets: Federal Investment in Research and Development at the Nation's Universities and Colleges. Santa Monica, CA: RAND. [Online.] Available: http://www.rand.org/publications/MR/MR1824/MR1824.sum.pdf [accessed May 20, 2004].
Fraumeni, B., and S. Okubo
2002 R&D in the National Income and Product Accounts: A First Look at Its Effect on GDP. [Online.] Available: http://bea.gov/bea/papers/R&D-NIPA.pdf [accessed April 1, 2004].
Gallaher, M.P.
2003 Measurement issues in a changing environment. Unpublished paper presented at the Panel on Research and Development Statistics at the National Science Foundation's Workshop on R&D Statistics at NSF, July 24-25, 2003, Washington, DC.
Gallaher, M., and J. Petrusa
2003 Technical Memorandum on Service Sector R&D. RTI Project No. 08236.002. Memorandum to the National Science Foundation. Research Triangle Park, NC: RTI International.
The Gallup Organization
2000 *Survey of Research and Development Funding and Performance by Nonprofit Organizations: Methodology Report*. Princeton, NJ: The Gallup Organization.
Gault, F.
2003 *Understanding Innovation in Canadian Industry*. Montreal: McGill-Queen's University Press.
Griffith, R., R. Harrison, and M. Hawkins
2003 Report on Estimating Private and Social Rates of Return to R&D Using Matched ARD and BERD Micro Data. Unpublished paper. London: Institute for Fiscal Studies.
Griliches, Z.
1980 *R&D and the Productivity Slowdown*. Working Paper 0434. Cambridge, MA: National Bureau of Economic Research.
Griliches, Z., and F. Lichtenberg
1984 Interindustry technology flows and productivity growth: A re-examination. *Review of Economics and Statistics* (May):324-329.
Groves, R.M., and J.M. Lepkowski
1985 Dual frame, mixed mode survey designs. *Journal of Official Statistics* 1:263-286.
Guellec, D., and B. Pattinson
2000 Innovation surveys: Lessons from OECD countries' experiences. *STI Review* 27:79.
Hall, B.
2001 Tax Incentives for Innovation in the United States. Report to the European Union submitted by Asesoria Industrial ZABALA-Spain. [Online.] Available: http://emlab.berkeley.edu/users/bhhall/papers/BHH01EUReportUSArtax.pdf [accessed April 12, 2004].

Hall, B.H., and W.F. Long
1999 Differences in Reported R&D Data on the NSF/Census RD-1 Form and the SEC 10K Form: A Micro-Data Investigation. [Online.] Available: http://emlab.berkeley.edu/users/bhhall/papers/HallLong99%20R&Ddata.pdf [accessed April 12, 2004].

Hansen, J.A.
2001 Technology innovation indicator surveys. In *Strategic Research Partnerships*, J.E. Jankowski, A.N. Link, and N.S. Vonortos, eds. Proceedings from an NSF Workshop. Washington, DC: National Science Foundation.

Hill, C.T., J.A. Hansen, and J.H. Maxwell
1982 Assessing the Feasibility of New Science and Technology Indicators. CPA-82-4. Boston: Center for Policy Alternatives, Massachusetts Institute of Technology.

Hill, C.T., J.A. Hansen, and J.I. Stein
1983 New Indicators of Industrial Innovation. CPA83-14. Boston: Center for Policy Alternatives, Massachusetts Institute of Technology.

Holmberg, A.
2002 Pre-Printing Effects in Official Statistics, An Experimental Study. Unpublished paper presented at the International Conference on Questionnaire Development, Evaluation, and Testing Methods, November 14-17, Charleston, SC.

Howells, J.
1999 Research and technology outsourcing. *Technology Analysis & Strategic Management* 11(March):17.

Jaffe, A.B
1996 Trends and patterns in research and development expenditures in the United States. *Proceedings of the National Academy of Sciences* 93:12658-12663.

Jankowski, J.
1999 Trends in academic research spending, alliances, and commercialization. *Journal of Technology Transfer* 24:55-68.
2002 Measurement and growth of R&D within the service economy. *Journal of Technology Transfer* 26:323.

Jones, A.K.
2003 Innovation in the U.S.—Descriptive Statistics. Unpublished paper presented at the Panel on Research and Development Statistics at the National Sciences Foundation's Workshop on R&D Statistics at NSF, July 24-25, 2003, Washington, DC.

Kish, L.
1965 *Survey Sampling*. Hoboken, NJ: John Wiley & Sons, Inc.

Koizumi, K.
2003 Understanding the Basic Federal Measures. Unpublished paper presented at the Panel on Research and Development Statistics at National Science Foundation's Workshop on R&D Statistics at NSF, July 24-25, 2003, Washington, DC.

Kott, P.S., and F.A. Vogel
1995 Multiple-frame business surveys. In *Business Survey Methods*, B. Cox, D.A. Binder, B.N. Chinnappa, A. Christianson, M.J. Colledge, and P.S. Kott, eds. New York: John Wiley & Sons, Inc.

Krugman, P.
2000 Where in the world is the "new economic geography"? In *The Oxford Handbook of Economic Geography*, G.L. Clark, M.P. Feldman, and M.S. Gertler, eds. Oxford: Oxford University Press.

Kuemmerle, W.
2003 Location of R&D Activity—International Data Issues. Unpublished paper presented at the Board on Science, Technology, and Economic Policy's Workshop on R&D Data Issues, April 7, Washington, DC.

Kuh, C.
2003 Does assessing doctoral programs bear any relation to classifying R&D? Presentation to the Board on Science, Technology, and Economic Policy's Workshop on Research and Development Data Needs, April 7, 2003, Washington, DC.

Kusch, G.L., and W. Ricciardi
1995 *Design of the Survey of Industrial Research and Development: A Historical Perspective.* Manufacturing and Construction Division Report Series, Working Paper Number Census/MCD/WP-95/01. Washington, DC: U.S. Census Bureau.

Larsson, A.
2004 Innovation output and barriers to innovation. In *Statistics in Focus*, KNS-NS-04-001-EN-N. Luxembourg: Eurostat.

Levin, R.C., A.K. Klevorick, R.R. Nelson, and S.G. Winter
1987 Appropriating returns from industrial research and development. *Brookings Papers on Economic Activity* 3:783-820.

Lichtenberg, F., and D. Siegel
1991 The impact of R&D investment on productivity: New evidence using linked R&D-LDR data. *Economic Inquiry* 29(2):203-229.

Link, A.N.
1996 On the classification of industrial R&D. *Research Policy* 25:397-401.
2003 University-Related Research Parks in the United States: Final Report. Paper submitted to the National Science Foundation, Research and Development Statistics Program. Department of Economics, University of North Carolina, Greensboro.

Long, W.F.
2003 Expanding the Collection of Data from Industry: Taking the Line of Business Step. Unpublished paper presented at the Panel on Research and Development Statistics at the National Science Foundation's Workshop on R&D Statistics at NSF, July 24-25, Washington, DC.

Mansfield, E.
1980 Comment on Griliches, "Returns to Research and Development Expenditures in the Private Sector." In *New Developments in Productivity Measurement and Analysis.* NBER Studies in Income and Wealth, Volume 44, J.W. Kendrick and B.N. Vaccara, eds. Chicago: University of Chicago Press.

McGuckin, R.H.
2004 *Internationally Comparable Science, Technology, and Competitiveness Indicators.* New York: The Conference Board.

Mohnen, P., and P. Therrien
2003 Comparing the innovation performance of manufacturing firms in Canada and selected European countries: An econometric analysis. In *Understanding Innovation in Canadian Industry*, F. Gault, ed. Montreal-Kingston: McGill-Queen's University Press.

Myers, S., and D.G. Marquis
1969 *Successful Industrial Innovations: A Study of Factors Underlying Innovation in Selected Firms.* NSF 69-17. Washington, DC: National Science Foundation.

National Research Council
1987 *Interdisciplinary Research in Mathematics, Science and Technology Education.* Committee on Research in Mathematics, Science, and Technology Education, Commission on Behavioral and Social Sciences and Education. Washington, DC: National Academy Press.
1995 *Allocating Federal Funds for Science and Technology.* Committee on Criteria for Federal Support of Research and Development, Commission on Physical Science, Mathematics, and Applications. Washington, DC: National Academy Press.

1997 *Industrial Research and Innovation Indicators: Report of a Workshop*, R.S. Cooper and S.A. Merrill, eds. Board on Science, Technology, and Economic Policy. Washington, DC: National Academy Press.

2000 *Measuring the Science and Engineering Enterprise: Priorities for the Division of Science Resources Studies*. Committee to Assess the Portfolio of the Division of Science Resources Studies of NSF. Washington, DC: National Academy Press.

2001a *Using Human Resource Data to Track Innovation: Summary of a Workshop*, S.A. Merrill and M. McGeary, eds. Board on Science, Technology, and Economic Policy. Washington, DC: National Academy Press.

2001b *Principles and Practices for a Federal Statistical Agency*, second edition. M.E. Martin, M.L. Straf, and C.F. Citro, eds. Committee on National Statistics. Washington, DC: National Academy Press.

2001c *Trends in Federal Support of Research and Graduate Education*. Committee on Trends in Federal Spending on Scientific and Engineering Research, Board on Science, Technology and Economic Policy. Washington, DC: National Academy Press.

2002 *Observations on the President's Fiscal Year 2003 Federal Science and Technology Budget*. Committee on the FY 2003 Federal Science and Technology Budget, Committee on Science, Engineering, and Public Policy. Washington, DC: The National Academies Press.

2003 *Improving the Design of the Scientists and Engineers Statistical Data System (SESTAT)*. Committee on National Statistics, Division of Behavioral and Social Sciences and Education. Washington, DC: The National Academies Press.

2005 *Facilitating Interdisciplinary Research*. Committee on Science, Education, and Public Policy. Washington, DC: The National Academies Press (forthcoming).

National Science and Technology Council

2003 *National Nanotechnology Initiative: Research and Development Supporting the Next Industrial Revolution*. Supplement to the President's FY 2004 Budget. [Online.] Available: http://www.nano.gov/html/res/fy04-pdf/fy04-main.html [accessed August 26, 2004].

National Science Board

2002 *Science and Engineering Indicators–2002*. [Online.] Available: http://www.nsf.gov/sbe/srs/seind02/toc.htm [accessed April 8, 2003.]

2004 *Science and Engineering Indicators—2004*. NSB 04-01. [Online.] Available: http://www.nsf.gov/sbe/srs/seind04/ [accessed July 8, 2004].

National Science Foundation

no date *Statistical Guidelines for Surveys and Publications*. Science Resource Statistics Division. [Online.] Available: http://www.nsf.gov/sbe/srs/infoqual/srsguide.pdf [accessed February 4, 2004].

1967 *Proceedings of a Conference in Technology Transfer and Innovation*. Washington, DC: National Science Foundation.

1976 *Technological Innovation and Federal Government Policy*. NSF 76-9. Washington, DC: National Science Foundation.

1996 *R&D Continues to be an Important Part of the Innovative Process*. Data Brief No. 7 (August 7). L.M. Rausch. Washington, DC: National Science Foundation.

1998 *Report on the NSF Agency Workshop on Federal R&D*. Washington, DC: National Science Foundation.

1999 *Report to the Congress: The Value and Usefulness of the RaDiUS Database*, May 20, 1999. Washington, DC: National Science Foundation.

2001 *Survey of R&D in Industry, 2001*, Table A-18. Washington, DC: National Science Foundation.

2002 *Methodology Report for the National Science Foundation's Survey of Federal Science and Engineering Support to Universities, Colleges, and Nonprofit Institutions, Fiscal Year 2000*. Washington, DC: National Science Foundation.
2004 *Largest Single-Year Decline in U.S. Industrial R&D Expenditures Reported for 2002*. Info Brief, NSF 04-320. Washington, DC: National Science Foundation.

Ohlsson, E.
1995 Coordination of samples using permanent random numbers. In *Business Survey Methods*, B. Cox, D.A. Binder, B.N. Chinnappa, A. Christianson, M.J. Colledge, and P.S. Kott, eds. New York: John Wiley & Sons, Inc.

ORC Macro
no date Revised Imputation Methodology for Basic Academic Research Data. Memorandum by Brandon Shackelford. Bethesda, MD: Quantum Research Corporation.
2002 Methodology Report for the NSF-NIH Survey of Scientific and Engineering Research Facilities, Fiscal Year 2001. Paper prepared for the National Science Foundation, September 2002.

Organisation for Economic Co-operation and Development
1963 *Proposed Standard Practice for Surveys of Research and Development: The Measurement of Scientific and Technical Activities*. Directorate for Scientific Affairs Report DAS/PD/62.47. Paris: Organisation for Economic Co-operation and Development.
1992 *Proposed Guidelines for Collecting and Interpreting Technological Innovation Data: Oslo Manual*. OECD/GD(92)26. Paris: Organisation for Economic Co-operation and Development.
2002a *Frascati Manual: Proposed Standard Practice for Surveys on Research and Experimental Development*. Paris: Organisation for Economic Co-operation and Development.
2002b *Purchasing Power Parities and Real Expenditures: 1999 Benchmark Year 2002 Edition*. Paris: Organisation for Economic Co-operation and Development.
2004 *Main Science and Technology Indicators*. Paris: Organisation for Economic Co-operation and Development.

Organisation for Economic Co-operation and Development and Eurostat
1997 *Proposed Guidelines for Collecting and Interpreting Technological Innovation Data: Oslo Manual*, second edition. Paris: Organisation for Economic Co-operation and Development.

President's Council of Advisors on Science and Technology
2002 *Assessing the U.S. R&D Investment*. [Online.] Available: http://www.ncseonline.org/affiliates/news/sep/PCAST_Report.pdf [accessed April 8, 2004].

Quantum Research Corporation
1999 *Study of Federally Funded Industrial R&D, Summary of Findings From Company Interviews and Analysis of Collateral Data*. Bethesda, MD: Quantum Research Corporation.

Reamer, A., L. Icerman, and J. Youtie
2003 Technology Transfer and Commercialization: Their Role in Economic Development. U.S. Economic Development Administration. [Online.] Available: http://www.eda.gov/ImageCache/EDAPublic/documents/pdfdocs/eda_5fttc_2epdf/v1/eda_5fttc.pdf [accessed April 7, 2004].

Romer, P.
2003 The soft revolution: Achieving growth by managing intangibles. In *Intangible Assets: Values, Measures, and Risks*, J.R.M. Hand, ed. Oxford, UK: Oxford University Press.

Rosenberg, N.
1994 *Exploring the Black Box: Technology, Economics and History*. Cambridge, UK: Cambridge University Press.
RR Bowker, Inc.
1994 *Directory of American Research and Technology*, 28th edition. New Providence, NJ: RR Bowker, Inc.
Schumpeter, J.
1947 *Capitalism, Socialism and Democracy*, second edition. New York: Harper and Row.
Statistics Canada
2003 *Industrial Research and Development: 2003 Intentions*. Catalogue no. 88-202-XIE. Ottawa, ON: Statistics Canada.
Street, D.L., N.B.Nichols, and S.J. Gray
2000 Segment disclosures under SFAS no. 131: Has business segment reporting improved? *Accounting Horizons* 14(3):259–285.
Tether, B., and S.J. Metcalfe
2002 Services and Systems of Innovation. Unpublished paper presented at the DRUID Academy Winter 2002 PhD Conference, Aalborg, Denmark, January.
Therrien, P., and P. Mohnen
2003 How innovative are Canadian firms compared to some European firms? A comparative look at innovation surveys. *Technovation* 23(4):359-369.
Thornton, P.H., and K.H. Flynn
2003 Entrepreneurship, networks, and geographies. In *Handbook of Entrepreneurship Research*, Z.J. Achs and D.B. Audretch, eds. Boston: Kluwer Academic Publishers.
Touhy, R.V.
1998 How the Department of Defense utilizes federal funds survey data. In *Report on the NSF Agency Workshop on Federal R&D*, National Science Foundation, ed. Washington, DC: National Science Foundation.
Tourangeau, R.
2003 Web-based data collection. In *Survey Automation: Report and Workshop Proceedings*, Oversight Committee for the Workshop on Survey Automation. D.L. Cork, M.L. Cohen, R. Groves, and W. Kalsbeek, eds. Committee on National Statistics. Washington, DC: The National Academies Press.
U.S. Bureau of Labor Statistics
1989 *The Impact of Research and Development on Productivity Growth*. BLS Bulletin 2331. Washington, DC: U.S. Bureau of Labor Statistics.
2002 *Multifactor Productivity Trends, 2000*. [Online.] Available: http://www.bls.gov/mfp [accessed October 8, 2004].
U.S. Census Bureau
1989 *Manufacturing Technology 1988*. Current Industrial Reports. Washington, DC: U.S. Census Bureau.
1994 Documentationof Nonsampling Error Issues in the Survey of Industrial Research and Development. Unpublished paper by D. Bond. Washington, DC: U.S. Census Bureau.
1995 Results of Cognitive Research on the Survey of Industrial Research and Development. Unpublished paper by W. Davis, and T.J. DeMaio. Washington, DC: U.S. Census Bureau.
1997 *Federally Funded Research and Development: Obligations Versus Spending, Interviews with Selected Performers*. Washington, DC: U.S. Census Bureau.

1998 Economic Data Collection via the Web: A Census Bureau Case Study. [Online.] Available: http://www.websurveyor.com/pdf/census.pdf [accessed February 4, 2004].

2000 *Data Collection Issues in the 1998 Survey of Industrial Research and Development.* Manufacturing and Construction Division, Special Studies Branch. Washington, DC: U.S. Census Bureau.

2004 2003 Survey of Industrial Research and Development, Form RD-1 Instructions. [Online.] Available: http://help.econ.census.gov/BHS/RD/RD-1I.pdf [accessed April 12, 2004].

U.S. Congress

2003 National Science Foundation Authorization Act of 2002, 107th Congress, 2d Session, H.R. 4664, Section 25. Study on Research and Development Funding Data Discrepancies.

U.S. Congress, House

1998 Unlocking Our Future: Toward a National Science Policy. Committee on Science. [Online.] Available: http://www.house.gov/science/science_policy_report.htm [accessed April 7, 2004].

U.S. Congress, Joint Economic Committee

1999a The Growing Importance of Industrial R&D to the U.S. Economy. [Online.] Available: http://garnet.acns.fsu.edu/~cskipton/JEC_post/r&d%20importance%20%20edwards.pdf [accessed June 3, 2004].

1999b American Leadership in the Innovation Economy. [Online.] Available: http://garnet.acns.fsu.edu/~cskipton/JEC_post/high-tech%20hearing%20report%201999.pdf [accessed June 3, 2004].

U.S. Department of Commerce

1967 *Technological Innovation: Its Environment and Management* (Charpie Report). Washington, DC: U.S. Department of Commerce.

U.S. General Accounting Office

2001 *Research and Development Funding: Reported Gap Between Data From Federal Agencies and Their R&D Performers Results From Noncomparable Data.* GAO-01-512R. Washington, DC: U.S. General Accounting Office.

U.S. Office of Management and Budget

2001 *Measuring and Reporting Sources of Error in Surveys.* Working Paper 31. Washington, DC: U.S. Office of Management and Budget.

2002 *Notice of Office of Management and Budget Action, Survey of Industrial Research and Development, 2001-2004.* Washington, DC: U.S. Office of Management and Budget.

2003a Circular No. A-11, Section 84-4. [Online.] Available: http://www.whitehouse.gov/omb/circulars/a11/current_year/s84.pdf [accessed April 8, 2004].

2003b *Statistical Programs of the United States Government, Fiscal Year 2004.* Washington, DC: U.S. Office of Management and Budget.

2003c *Implementation Guidance for the E-Government Act of 2002.* Memorandum. Washington, DC: U.S. Office of Management and Budget.

2004 *President's Budget for 2005: Analytical Perspectives, Research and Development.* [Online.] Available: http://www.whitehouse.gov/omb/budget/fy2005/pdf/spec.pdf [accessed May 20, 2004].

Vonortas, N.

2003 Data on R&D Collaboration. Unpublished paper presented at the Board on Science, Technology, and Economic Policy's Workshop on R&D Data Needs, April 7, Washington, DC.

Appendix A

Workshop Agenda

Workshop on Measurement of Research and Development

July 24-25, 2003

Open Meeting

July 24, 2003 Location: Room 100, 500 Fifth Street, N.W., Washington, DC 20001

7:30-8:15 Continental Breakfast

8:15-8:30 Welcome and Opening Remarks
Larry Brown

8:30-9:00 Congressional Uses of R&D Data
David Goldston
Majority Chief of Staff
House Science Committee

9:00-9:30 National Science Board Use of R&D Data
Anita K. Jones
National Science Board

9:30-10:30 Panel 1: Refining the Measures of R&D
John Adams
Moderator

Basic, Applied, Development: Do We Need New Boundaries
Greg Tassey
NIST

Measurement Issues in a Changing Environment
Michael Gallaher
RTI

10:30-10:45 Break

10:45-12:00 Panel 2: Statistical Methodology Issues
Richard Valliant
Moderator

Developing a Quality Profile of the NSF Portfolio
Barbara Bailar
Consultant

Improving Survey Methodology
Jeri Mulrow
NSF

Research Agenda
Ray Wolfe
NSF

12:00-1:00 Lunch

1:00-3:00 Panel 3: Improving Measures of R&D Activity in Industry
Steve Klepper
Moderator

Census Bureau Perspective on Improving the RD-1 Survey
Bill Bostic
Census Bureau

Expanding the Collection of Data from Industry
Bill Long
BPRA Inc.

Collecting R&D Data from Businesses in Japan
Tomohiro Ijichi
University of Tokyo

Producing Geographic Estimates
Michael Bordt
Statistics Canada

3:00-3:30 Break

3:30-4:30 Panel 4: Emerging Uses of R&D Data
Bronwyn Hall
Moderator

R&D Data in the National Accounts
Barbara Fraumeni
BEA

July 25, 2003 Location: Room 101, 500 Fifth Street, N.W., Washington, DC 20001

8:00-8:30 Continental Breakfast

8:30-9:30 Panel 5: Measures of R&D Spending in the Federal Government
Jay Hakes
Moderator

Understanding the Basic Federal Measures
Kei Koizumi
AAAS

Supplementing the NSF Surveys: Lessons
Donna Fossum

Learned from the RaDiUS Program
RAND

9:30-9:45 Break

9:45-11:00 Panel 6: Obtaining Data on Innovation
Wes Cohen
Moderator

Surveys of Innovation
Mike McGeary
Consultant

European Community Innovation Survey (CIS-3)
Anna Larsson
(via teleconference)
Sverre Dommersnes
EUROSTAT

Lessons Learned in the Carnegie-Mellon Survey
John Walsh
University of Tokyo

11:00-11:30 Workshop Summary and Discussion
Larry Brown, Bronwyn Hall

11:30-12:30 Lunch

Appendix B

Biographical Sketches of Panel Members and Staff

Lawrence D. Brown (*Chair*) is the Miers Busch professor in the Department of Statistics at the Wharton School of the University of Pennsylvania. Previously, he has held faculty positions at Cornell University, Rutgers University, and the University of California, Berkeley. A theoretical statistician, his primary area of research interest is mathematical statistics, in particular, statistical decision theory, complete class theorems, admissibility and minimaxity of estimators and tests, sequential analysis, and the foundations of statistics (properties of conditionality and ancillarity). He was elected to membership in the National Academy of Sciences in 1990. At the National Research Council (NRC), he is currently a member of the Panel to Review the 2000 Census. He has a B.S. from the California Institute of Technology and a Ph.D. from Cornell University.

John L. Adams is senior statistician at the RAND Corporation in Santa Monica, California. His recent research areas include measurement of the quality of health care in view of outcomes and construction and evaluation of simulation models. He is a visiting professor in applied times series analysis and data collection, analysis, and presentation topics. He was a member of the NRC Panel on Data and Methods for Measuring the Effects of Changes in Social Welfare Programs. He has a Ph.D. in statistics from the University of Minnesota.

Barbara A. Bailar (*Consultant*) is a statistical consultant in Washington D.C. From 1995 to 2001, she served as Senior Vice-President for Survey Research at the National Opinion Research Center. From 1988 to 1995, she was the Executive Director of the American Statistical Association. Prior to that, she was the Associate Director for Statistical Standards and Methodology at the Bureau of the Census. She is a Fellow of the American Statistical Association and the American Association for the Advancement of Science. She is a past President of the American Statistical Association and of the International Association of Survey Sampling, and a former Vice-President of the International Statistical Institute. She has published many articles on censuses and surveys in a variety of publications.

Wesley M. Cohen is professor of economics and management at the Fuqua School of Business at Duke University, as well as a research associate at the National Bureau of Economic Research. He has also served as an economist at the Energy Information Agency of the U.S. Department of Energy. His research interests focus on the economics of technological change and industrial organization economics. His published work explores many facets of industrial research and development policy and the economic roots of innovation, and he utilizes the NSF R&D data extensively in his research and writing. He has a Ph.D. in economics from Yale University (1981).

Fred Gault is the director of the Science, Innovation and Electronic Information Division at Statistics Canada, where he is responsible for the development of statistics on all aspects of research, development, invention, innovation, and diffusion of technologies. He is a fellow of the Institute of Physics and a member of the British Computer Society. He is chairman of the Committee of National Experts on Science and Technology Indicators at the Organisation for Economic Co-operation and Development and was chairman of the Working Party on Indicators for the Information Society at the Organization for Economic Co-operation and Development from its beginning in 1997 until 2002. He has a B.Sc. in economics and a Ph.D. in mathematical physics from the University of London.

Marisa A. Gerstein is a research assistant with the Committee on National Statistics. She has worked on numerous projects, including panels on elder mistreatment, nonmarket accounts, research and development statistics, and the 2000 and 2010 decennial censuses. She has a B.A. in sociology from the New College of Florida.

Jay Hakes is director of the Jimmy Carter Library and Museum in Atlanta. From 1993 to 2000 he served as an administrator of the Energy Information Administration in the U.S. Department of Energy. In that position, he oversaw the collection, dissemination, and publication of the national energy data series. His areas of interest include statistical survey management, strategic planning, and data dissemination. He has a Ph.D. from Duke University.

Bronwyn H. Hall is professor of economics in the Department of Economics at the University of California, Berkeley, as well as a research associate at the National Bureau of Economic Research and president of TSP International, a computer software firm. Her research has focused on financing R&D, R&D in academic settings, and innovation in the economy. She has used NSF R&D data in comparative studies with other sources, such as patent counts. She has extensive service on NRC boards and panels and currently is a member of the Board on Science, Technology, and Economic Policy; she is also chair of its Planning Committee for Workshop to Review Research and Development Statistics at the National Science Foundation, the work of which is being coordinated with this committee. She has a Ph.D. in economics from Stanford University (1988).

Christopher T. Hill is vice provost for research and professor of public policy and technology at George Mason University. He is also president of George Mason Intellectual Properties, Inc. He has served on the professional staff at the National Academy of Engineering, the National Research Council, and the Congressional Research Service. He is a fellow of the American Academy of Arts and Science. His publications have been in the field of technological innovation and its impact on the economy, the impact of federal regulation on innovation, and the university perspective on issues of federal R&D procurement. He is responsible for completion of the NSF survey of academic R&D at George Mason University. He has a Ph.D. in chemical engineering from the University of Wisconsin.

Steven Klepper is professor of economics and social science in the Department of Social and Decision Sciences at Carnegie Mellon University. He is an affiliate of the H. John Heinz III School of Public Policy and Management at Carnegie Mellon. His fields of specialization include the evolution of industry and the determinants of technological change, statistical procedures to cope with measurement error, and tax compliance. He is on the editorial board of *Law and Society Review* and the *Journal of Criminal Law and Criminology*. He is a research associate in the Centre for Research on Innovation and Competition, University of Manchester, and has served on the economics panel of the National Science Foundation. His books and articles focus on innovation, economic development, economic evolution, and technological change as a factor in growth and decline of industry. He has a Ph.D. in economics from Cornell University (1975).

Tanya M. Lee is a project assistant at the Committee on National Statistics. Previously at the Institute of Medicine she worked on the Committee on Strategies for Small Number Participant Clinical Research Trials and the Committee on Creating a Vision for Space Medicine During Travel Beyond Earth Orbit. She is pursuing a degree in the field of psychology.

Joshua Lerner is the Jacob H. Schiff professor of investment banking at the Harvard Business School with a joint appointment in the finance and entrepreneurial management units, as well as a research fellow at the National Bureau of Economic Research. He has extensive experience on issues concerning technological innovation and public policy at the Brookings Institution and on Capitol Hill. His research focuses on the structure of venture capital organizations and issues of patents and other intellectual property protection on the competitive strategies of high-technology industries. He served on the NRC Committee on the Workshop on U.S.-Japan Technology Links in Biotechnology. He has a Ph.D. in economics from Harvard University.

Baruch Lev is professor of accounting and finance at the Stern School of Business of New York University and director of the Vincent C. Ross Institute for Accounting Research and the Project for Research on Intangibles. He is a member of the editorial boards of the *Journal of Accounting Research* and the *Journal of Accounting and Public Policy*. His work focuses on understanding the measurement, valuation, and reporting issues concerning intangible investments, among which research and development are a primary classification. He is also interested in the impact of accounting rules, particularly FASB Statement 2, on the organization and reporting of R&D. He has a Ph.D. from the University of Chicago (1968).

Gary McDonald recently retired from the General Motors Research & Development Center, where he was appointed head of the Mathematics Department in 1983; head of the Operations Research Department in 1992; and director of the Enterprise Systems Laboratory in 1998. He is currently a visiting professor at Oakland University and an assistant director at the National Institute of Statistical Sciences. He has chaired several NRC panels and has published numerous articles in diverse areas of applied and mathematical statistics. He is a fellow of the American Statistical Association, the Institute of Mathematical Statistics, and the American Association for the Advancement of Science. He has a B.A. from St. Mary's University of Minnesota and M.S. and Ph.D. degrees in mathematical statistics from Purdue University; he received a doctor of science *honoris causa* from Purdue University in 2000.

Michael McGeary (*Consultant*) is a political scientist who directed the staff work for a dozen reports by committees of the Institute of Medicine (IOM) and other units of the National Academies.

Thomas J. Plewes (*Study Director*) is a senior program officer for the Committee on National Statistics. In addition to the Committee for the Review of Research and Development Statistics at the National Science Foundation, he is directing studies of international trade traffic statistics and supporting NRC initiatives with the U.S. General Accounting Office on key national indicators of performance. Prior to joining the NRC staff, he was associate commissioner for employment and unemployment statistics of the Bureau of Labor Statistics. He is a fellow of the American Statistical Association and was a member of the Federal Committee on Statistical Methodology. He has a B.A. degree from Hope College and an M.A. degree from the George Washington University.

Nora Cate Schaeffer is a professor in the Department of Sociology at the University of Wisconsin, Madison. Her research involves cognitive aspects of survey questionnaire design and evaluation of interviewer effects on survey responses. She has served as a visiting fellow at the Census Bureau's Center for Survey Methods Research. She is a member of the Committee on National Statistics and has served as a member of its Panel to Evaluate Alternative Census Methods. She has a Ph.D. in sociology from the University of Chicago (1984).

Richard Valliant is senior research professor at the University of Michigan and professor in the Joint Program in Statistical Methodology at the University of Maryland. He has over 25 years of experience in sample design and estimation using data from complex surveys. He is currently an associate editor of *Survey Methodology* and the *Journal of Official Statistics* and has also been an associate editor of the *Journal of the American Statistical Association*. He is a fellow of the American Statistical Association. He has a Ph.D. in biostatistics from Johns Hopkins University (1983).